Volcanoes
and the
Making *of* Scotland

Volcanoes
and the
Making *of* Scotland

Brian Upton

Dunedin Academic Press

EDINBURGH

Published by
Dunedin Academic Press Ltd
Hudson House
8 Albany Street
Edinburgh EH1 3QB
Scotland

ISBN 1 903765 40 4

British Library Cataloguing in Publication Data
A catalogue record for this book is available from the British Library.

*Front cover and frontispiece: A thick succession of
Palaeogene lavas, Ben More, Mull. Picture by Brian Upton.*

Designed and typeset by David McLeod,
Makar Publishing Production, Edinburgh.
Printed in Great Britain by Scotprint, Haddington.

Contents

Poetry in stone! A natural staircase of basalt lava, Staffa.

Foreword

There have been so many times during my geological travels in Scotland when I have met farmers, fishermen, tourists and others who have been curious about what I was doing and who have asked me to explain, in simple terms, the origin of the rocks and landscape features around them. Having spent a professional lifetime publishing papers for scientific readership as well as teaching, almost exclusively, to geology students in one of Scotland's oldest universities, I felt it time to attempt to write a reasonably non-technical explanation for the general public. My own work, which has been devoted to volcanology and igneous petrology over the past fifty years, has afforded me the privilege of working in areas of active or recent volcanism as far apart as Iceland, East Africa, the western Indian Ocean and the United States.

While Scotland may lack active volcanoes, its landscape is crowded with the ruins of volcanoes that erupted in bygone ages. Many of its well-known topographic features represent the root zones of volcanoes, the upper parts of which have been shaved off by erosion. This is true for some of the well-known topographic features in central Scotland such as Dumbarton Rock near Glasgow and Castle Rock, Edinburgh, as well as for many of the most scenic parts of the Hebrides and western Highlands. In general, the rocks from which the Scottish landscape has been sculpted record an extremely long and complex history of events in which volcanism has played a major role.

Had this account been intended as a scientific treatise it would be littered with references to relevant publications and there would be a long listing of the all those publications that I have consulted. But it is not and I refer only to a selection of sources that provide a lead to those readers looking for more detailed information. In writing this book I have helped myself very freely to material from the published literature. I am conscious of the huge debt of gratitude I owe to all who have taught me about the igneous rocks of Scotland. Without the encouragement of Gordon Craig and Douglas Grant the book would never have started. Among those others who have given specific help my sincere thanks go to Brian Bell, Euan Clarkson, Henry Emeleus, Angus

Harkness, Ella Høch, Graham Leslie, Alastair Robertson, Jack Soper and David Stephenson for their time and wisdom. To any who may feel that I have misrepresented their ideas or conclusions, I ask forgiveness. It is difficult to find appropriate words of thanks for my wife who has shown the utmost forbearance and patience while the book was being written.

.

Chapter 1

Introduction

Scotland today presents a peaceful landscape. Earthquakes are rare and mercifully small and the nearest active volcanoes lie far to the north-west in Iceland, or away to the south-east in Italy. Admittedly we do not have to travel quite so far to see landscapes that are easily recognisable as volcanic; in central France the Chaîne des Puys consists of volcanic hills that have been little modified by erosion since they last erupted several tens to hundreds of thousands of years ago. The Romans, who recognised them as old volcanoes were, of course, thoroughly familiar with such features at home. However, a record of volcanism over an immense period of time is preserved in the Scottish rocks, and this book attempts to present a guide for amateur geologists to Scotland's dramatic and fiery past. Evidence of the former existence of volcanoes, in more or less ruinous and fragmentary state, is scattered in a thousand sea-cliffs, road-cuts, quarries, hillsides and mountains from the Shetlands to the English border.

While there is general public awareness that Scottish landscape features such as Arthur's Seat, Staffa, Glen Coe and the Cuillin are of volcanic origin, there is a conceptual difficulty in relating what one actually sees to the perception of an active volcano. Terminology is one of the main stumbling blocks. Lack of understanding of the relationship between intrusive igneous rock bodies formed within and beneath volcanoes and the extrusive phenomena with which they may have been associated represents another. This book is written as a non-technical account of the volcanic history of Scotland.

In general, the older the suite of volcanic rocks the more the accidents and crises throughout geological time (e.g. erosion, faulting and folding) have made the obvious links between an outcrop of old volcanic rocks and a real volcano ever more tenuous. Consequently, whereas it is traditional in geology texts to start with the oldest rocks and work 'upwards' towards younger formations, I am deliberately reversing this convention. I shall start with the youngest rocks and go back in time to features in the oldest rock formations in Scotland, which date from roughly three thousand million years ago. It is easier to understand how features such as the Cuillin, a mere sixty million years old, might represent the sawn-down ruins of a once great volcano than it is to try to do the same with, for example, a much older and highly contorted suite of volcanic rocks such as are seen on the south-west coast near Ballantrae. It is consequently my intention to invite the reader to go backwards through time to consider the origins of the remarkable range and variety of volcanic rocks whose outcrops make up much of the Scottish landscape.

This account of the ancient volcanoes is intended neither as an academic treatise nor as a field guide. My choice of volcanoes is to a large extent idiosyncratic while, at the same time, including the better known topographic features – such as Ben Nevis, the Cuillin, Arthur's Seat or Dumbarton Rock – whose rocks had a volcanic origin.

It is necessary first to define what a volcano is and then give some background as to how and why a volcano happens. A volcano marks the place where molten material is vented from the interior of the Earth onto the surface. The molten material is called magma. Release of gas may pre-date and/or post-date the eruption of magma, and invariably accompanies its eruption. Eruption of magma may take place through a conduit which may be roughly pipe-like or one that has the form of an elongated crack or fissure. Hence volcanoes may be crudely subdivided into those of 'central-type', where the plan is roughly circular, and those of 'fissure-type', where the plan is elongate. The eruptive products are gases, lavas (which flow out) and fragmental materials (which are blown out). The gases mainly consist of water and carbon dioxide (but with many other components). At depth the gases are held in solution in the magma. If we consider a champagne bottle, the gas (carbon dioxide) is kept in solution for as long as the pressure in the bottle is maintained. Uncorking the bottle, thereby releasing the pressure, allows the gas to come out of solution to produce the bubbles and froth. We may use this analogy to appreciate that gases (or potentially volatile materials) kept in solution in the magma while the pressure remains sufficiently high, will separate from the melt in near-surface, low-pressure environments.

The lavas represent the congealed magmas that have been degassed to varying degrees. The fragmental deposits may be exclusively derived from the magma, dispersed into particles or droplets, or from the solid rocks that form the sides of the magma conduits (so-called 'country-rocks') that have been violently broken up by the explosive release of escaping gas. Commonly both sources contribute to the fragmental deposits, conveniently termed 'pyroclastic' deposits, from the Greek stem meaning 'fire-broken'. The fragments or particles themselves are then 'pyroclasts'. The terms 'volcanic ash' and 'cinders' still retain wide currency, although the idea of volcanoes producing ash or cinders dates back more than two hundred years to when it was believed that volcanoes were the result of underground fires. Where magma reaches the surface, piles of lava and pyroclasts may build up, sometimes forming volcanic mountains rising to heights of as much as 6 or 7 km above sea level. If the magma conduit was essentially cylindrical a conical heap of products will result. *Fig. 1.1* depicts a simple central-type cone topped by a crater. Mt Egmont in New Zealand (*Fig. 1.2*) is a fine example of such a central-type volcano, conforming closely to the popular idea of what a volcano *ought* to look like.

If there are several feeders a composite multi-vent volcano will be formed. If, as is common, large volumes of magma are stored at shallow levels beneath or within the volcanic edifice, their sudden emptying in a major eruption or by any other mechanism may result in collapse of the unsupported overlying pile, truncating the volcano and producing a large pit called a caldera (*Fig. 1.3*). The diameter of these calderas is believed to roughly equate with that of the magma body (reservoir or chamber) that preceded them; diameters of up to 10 km are commonplace although calderas several times larger are known on Earth.

If, rather than having a more or less identifiable focal point of eruption as in a central-type volcano, the conduit is an elongate split, a fissure volcano can result (*Figs 1.4 and 1.5*).

Few sub-aerial fissure eruptions have occurred in historic times, and these have been almost exclusively in Iceland. Some of the most recent fissure

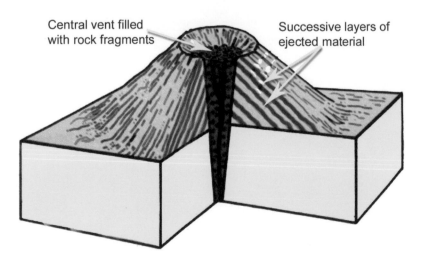

Fig. 1.1 A simple central-type volcano with a crater and downward narrowing conduit. (*After F. Press and R. Siever, 1982*)

Above, Fig. 1.2 Mt Egmont, a large central-type volcanic cone in North Island, New Zealand. *Below, Fig. 1.3* Several nested collapsed caldera pits on Kartala Volcano, Grande Comore Island, Mozambique channel. White vapour rises from a small ash-cone within the caldera.

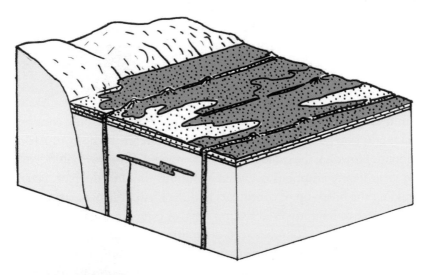

Fig. 1.4 Block diagram illustrating magma rising along elongate tensional splits forming dykes. Magma reaching the surface erupts in fissure volcanoes. *(After R.S. Fiske, 1971)*

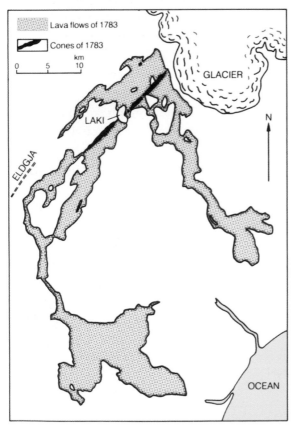

Fig. 1.5 Map of Laki volcano, southern Iceland. Black indicates the fissure zone: yellow is the basalt lava erupted from it in 1783. *(After R. Decker and B. Decker, 1983)*

eruptions occurred at Krafla Volcano, northern Iceland, between 1975 and 1983, when opening of linear fissures, up to a kilometre long, allowed very mobile, incandescent basalt magma to fountain up as 'curtains of fire' attaining heights of several hundred metres (*Fig. 1.6*).

The largest historically recorded fissure eruption occurred in 1783 when a fissure some 25 km long developed at Laki in SE Iceland (*Fig. 1.5*). Apart from these eruptions on the oceanic island of Iceland, the only continental fissure eruption in historical times is a small one in the horn of Africa in Djibouti. However, the geological record provides incontrovertible evidence that great fissure eruptions occurred at various times in the past, generally heralding the break-up of continents and the birth of new oceans. Some of these appear to have involved fissures tens, or even hundreds, of kilometres in extent. The distinction between central-type and fissure-type volcanoes is useful but, as so often with natural phenomena, all intermediate varieties can be encountered, and a single volcanic system may during its active life change from one to the other.

Some of the steepest-sided volcanoes are those whose products are predominantly fragmental. The stable slopes (or angles of rest) of these are commonly around 35° to the horizontal, much as in ash-piles or slag-heaps. If the fragments become cemented into coherent material they form pyroclastic rocks. For those volcanoes where lavas predominate over pyroclastic materials, the angle of slope is commonly defined by the fluidity of the lavas. Extremely viscous lavas, which may have flow properties more akin to pitch or tooth-paste, flow thickly and slowly and do not travel far from the vent from which they were erupted. Steep-sided domes or spines of such tacky lava will result.

Basalt lavas, which are far and away the most abundant variety, can be erupted at temperatures of well over 1,000° C. At these high temperatures they are very mobile and flow and spread readily. Huge basaltic eruptions in ancient times appear to have flowed for distances of over 100 km from their vents with such low angles of slope as to be virtually horizontal. In brief, the subject of volcano-morphology is complex and volcanoes can present many forms, often differing strikingly from the commonly held image of a simple cone surmounted by a relatively small vent. Furthermore, active volcanoes should be thought of as dynamic entities whose shape may change greatly during their evolution. Of the various factors controlling their geometry, two of the most important are the relative ratio of pyroclastic deposits to lavas and the viscosity of the lavas.

Fig. 1.6 Lava fountaining during a fissure eruption at Krafla.

The constitution of the Earth and the nature of tectonic plates

Before attempting any account of Scotland's old volcanoes, an overview of the Earth's structure and the subject of plate tectonics is called for to gain some understanding of how volcanoes come about and why they occur where they do.

The distance from the surface of the Earth to its centre is about 6,300 km (*Fig. 1.7*). Getting on for half-way down, at around 2,900 km, we know from a whole series of indirect observations (mainly on the behaviour of earthquake waves) that there is a dramatic change in the physical and chemical nature of the materials. This defines the break between the enveloping mantle and the central core. The core is inferred to be predominantly composed of an iron–nickel metallic alloy whereas the surrounding mantle is mainly composed of silicate rock. Although the composition of the mantle does not change dramatically from the core boundary outwards, it does have some notable changes in its physical properties, of which the most important are changes in its density, which reduces in several steps on approaching the surface. In its outermost density shell (about 650 km thick) the rock is mainly composed of the mineral olivine and is known as peridotite, derived from the French word *peridot*, meaning olivine.

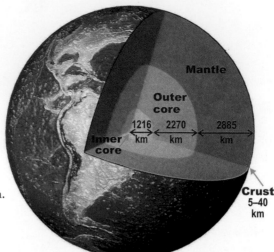

Fig. 1.7 Cut-away diagram showing the internal structure of the Earth.

Although rocks appear hard and brittle, they have a Jekyll and Hyde nature; given sufficient time, in conjunction with the application of heat and pressure, they are capable of flow and exhibit some of the properties of a fluid. In contrast, in response to short-term stress (for example, a blow with a hammer!) they will exhibit their familiar brittle behaviour and will fracture. A well-known illustration of this dual behaviour can be found in ice. An icicle is hard and brittle and will snap like a twig. But ice can also flow, as we know from the existence of glaciers. A more dramatic demonstration of a material which, according to the rate at which stresses are applied, changes its character is the silicone material known as 'bouncing putty'. Thrown on the floor it lives up to its name by bouncing like an elastic solid, whereas a lump left in a saucer for a while will collapse and flow out like melted wax. All of this is relevant because there is now ample evidence that the rocks composing the mantle are not static but, despite staying solid, are in continuous slow motion. The idea of rocks being solid but at the same time capable of flow may take some time to get used to!

Small quantities of radioactive elements (for example uranium) are present in the mantle but they are irregularly distributed. The spontaneous decay of such unstable radioactive atoms produces heat. The irregular distribution of these heat-producing elements produces, over long periods of time, localised hot-spots in the mantle. Consequent thermal expansion reduces the density, i.e. it increases the buoyancy, with consequent slow ascent of the hot rock through solid-state flow. At the same time, cold and dense portions of rock in the outer zones of the Earth sink deep into the mantle to participate in an age-old pattern of convective overturn.

Whereas rocks of the deep mantle are confined under such high pressures that melting does not occur, those of the outer part (from a few tens to a hundred kilometres down) and held under lesser pressures, are at temperatures only slightly below (and sometimes above) those at which melting commences. The arrival of hot ascending mantle rock, in the form of what are generally referred to as 'thermal plumes', beneath the outermost shell of the Earth is extremely effective in promoting melting in the upper mantle, forming the magmas that supply volcanoes. At the present day, such thermal (or mantle) plumes are widely believed to provide the energy for volcanism in Iceland and Hawaii and probably that in the Canaries and Azores islands. As will be shown, ancient mantle plumes were probably responsible for some of the volcanoes whose products we can see in Scotland today.

The non-convecting outer shell of the solid Earth is called the lithosphere. This shell, which varies in thickness from a few kilometres to one or two hundred kilometres, comprises both the outermost mantle portion and the overlying crust. There are major mineralogical differences between the rocks forming the crust and those forming the denser mantle beneath: olivine is a very minor component in crustal rocks in contrast to the underlying mantle peridotites, in which it predominates. The principal minerals in the crust are the alumino-silicates of the feldspar group, which are, to all intents and purposes, absent in the mantle. In terms of elements, the feldspars are built up of oxygen, silicon and aluminium together with lesser amounts of calcium, sodium and potassium. As a further complication, the crustal rocks of the oceanic regions differ chemically and mineralogically from those of the continental regions. The continental crust, especially in its uppermost 10–15 km, is much richer in silicon, sodium and potassium than oceanic crust, and this chemical difference is reflected in the rock types, which are largely granites and allied rocks. Furthermore, the continental crust is thicker than that beneath the oceans. Typical thicknesses for continental and oceanic crust are around 30 and 11 km, respectively.

The lithosphere, however, does not represent a continuous, mechanically-coherent world-encircling shell but is subdivided into irregular polygonal sections of variable size, referred to as tectonic plates. These can be considered as passive slices of rock 'floating' on the more mobile (but still solid!) convecting interior described above. Boundaries between one tectonic plate and its neighbours are of three types.

The first are called *transform boundaries*, at which the plates may move laterally past one another, with the boundary plane between them constituting a fault of the type known as a transform fault (*Fig. 1.8*). Rather than the plate motions being smooth and continuous, they are typically jerky. The plates tend to interlock and stick until stresses built up by flow in the underlying mantle reach a point where the plates break free and move violently, sometimes by as much as several metres. The release of accumulated strain energy, when the adjacent plates do move, produces earthquakes, sometimes of prodigious destructiveness. A famous active example is the San Andreas Fault of California, whereas the Great Glen in Scotland marks the trace of an ancient transform fault of this kind. Such transform plate boundaries are not, however, typically associated with volcanism.

The second category of plate boundary is that where the two plates pull away from one another as if to create void space between them (*Fig. 1.8*). Void space is, however, not generated because underlying (solid) mantle peridotite wells up between them. In the resulting extensional environment the peridotite beneath the thinned stretched lithosphere starts to melt. This process, and the other ways in which the mantle can become partially molten are explained in *Figure 1.9*.

The magma thus produced ascends (because of its relatively low density) and then crystallises to rocks composing new oceanic lithosphere. Such extensional plate boundaries almost invariably occur in the oceans, along the crest of 'mid-ocean ridges'. In the Atlantic Ocean, the Red Sea and the Gulf of Aden the plate boundary lives up to its name in forming a 'mid-ocean ridge' virtually equidistant from the continental margins. However, in the Pacific and Indian Oceans the submarine volcanic ridges marking these extensional plate boundaries are distributed more eccentrically. Melting below these extensional plate boundaries, in which some 10–15% of the underlying peridotite is consumed through melting, produces the commonest type of magma, namely basalt magma. The 85–90% of the peridotite that is left behind, i.e. the refractory residue deprived of its more fusible basaltic components, is incorporated into new oceanic lithospheric mantle. Some of the basalt magma reaches the ocean floor to form submarine volcanoes, whose eruptions are rarely recorded, let alone seen. Much of the magma, however, crystallises at depth, forming intrusive rocks. In summary, the new oceanic lithosphere created by these extensional plate boundaries comprises a lower layer of residual (refractory) peridotite overlain by crust composed of intrusions and extrusive lavas generated by the cooling magma derived from the mantle. Because of the generation along them of new oceanic lithosphere such plate boundaries may alternatively be called *constructional plate boundaries (Fig. 1.8)*.

It is by the persistent activity beneath and along the interconnecting 70 000 km of the planet's mid-ocean ridges, over tens of millions of years,

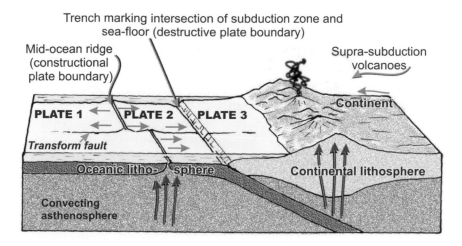

Fig. 1.8 The relationship of different types of plate boundaries. Arrows arising from the convecting asthenosphere and the subducting plate indicate ascending magama.

that new oceanic floor forms in a more or less continuous fashion. The opposing sides of these constructional plate boundaries move apart at rates of anything up to about 15 cm per year. As with transform boundaries, the movement is not smooth but spasmodic. Extension along one sector, which may be tens of kilometres long, involving both volcanism and seismicity, may be followed by decades or centuries of inactivity before another sector, perhaps hundreds of kilometres further along the boundary, takes up the action. An average separation of, let us say, 10 cm per annum is thus achieved in a jerky or jumpy manner rather than in the continuous manner of a moving walk-way.

It is not too fanciful to consider the whole of the mid-ocean ridge system as constituting one vastly elongate fissure volcano, although eruptions localised along its length may be widely separated both in space and time. Whereas mid-ocean ridge volcanism is rarely observed in action and the present-day ocean floor so formed is, by its nature, generally unavailable for inspection, we can nevertheless see some of its ancient products exposed in Scotland.

Clearly, if there were only transform and constructional plate boundaries, the lithosphere would continue to expand indefinitely. Consequently the third type of boundary, along which lithospheric plates are consumed and recycled deep in the mantle, plays an essential role in the process. In these *destructive plate boundaries* old oceanic crust (where 'old' can mean anything up to around 200 million years) eventually sinks back into the depths of the mantle (*Fig. 1.8*). The process is called subduction and occurs because, as the oceanic lithosphere ages, it cools and thickens and acquires a density greater than that of the underlying mantle peridotites so that, like a submarine that has blown its tanks, it begins its descent to the deep. Again one must bear in mind the fact that the mantle rocks are not molten but are solids that are capable of flow in response to stresses applied over long periods of time by a process of continual recrystallisation as in the glacier simile. The descending plate may take a near vertical dive but more commonly descends at a shallower angle (as depicted in *Fig. 1.8*), dipping away from the mid-ocean ridge where it was first formed. The slice of mantle above such an obliquely descending oceanic plate, clearly triangular in cross-section, is referred to as the overlying 'mantle wedge'.

Let us return to the mid-ocean ridges. As new crust grows, fractures and fissures develop in it that allow sea-water to percolate down to encounter ever hotter rocks. The water reacts with these hot rocks forming assemblages of hydrated minerals ('hydrothermal reactions'). Consequently, when eventually (tens or even hundreds of millions of years later) the oceanic lithosphere begins its descent, it carries with it a large amount of water chemically bonded into it. As it plunges deeper into the mantle it experiences rising pressures and temperatures and a complex sequence of dehydration reactions takes place releasing the water that had previously been incorporated. The water thus liberated (together with components in solution) rises into and reacts with the mantle wedge above. Now water has the property of reducing the melting point of peridotite, thus the introduction of water renders the peridotite (*Fig. 1.9*) within the mantle wedge more fusible. This, in conjunction with the frictional heat released through the abrasion of the descending plate against the adjacent overlying mantle wedge, causes partial melting in the wedge. The magmas thereby generated rise above the sinking plate and erupt at the surface in what are called supra-subduction volcanoes. Typically volcanoes of this kind form chains some 100 to 200 km back from the line along which the descent of the plate begins (*Fig. 1.8*).

The intersection of dipping slices or slabs of oceanic mantle with the subspherical surface of the Earth produces arcuate linear features as would, for example, be made by oblique knife-cuts on the surface of an orange. The arcuate feature where descent commences is normally a deep ocean trench (*Fig. 1.8*), whereas the supra-subduction magmatism typically forms a volcanic arc that is convex towards the ocean. A glance at a map of e.g. the Pacific Ocean reveals a profusion of such trenches and associated volcanic arcs (particularly in the north and west, where the Aleutian and Marianas represent arcuate chains of volcanic islands), formed respectively along and

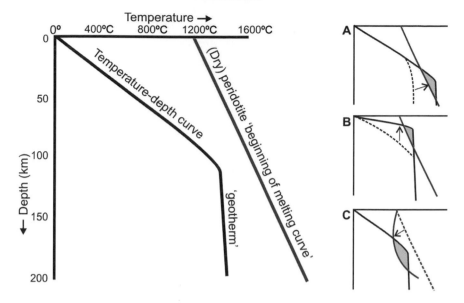

Fig. 1.9 Change in temperature with depth in the Earth and illustrations of the three means by which mantle peridotites begin melting to produce basaltic magmas.

The black curve (technically known as the 'geotherm') in the left-hand diagram shows how temperature increases with depth in the Earth. At first the temperatures rise sharply with increasing depth but, at greater depths, the curve steepens abruptly so that further increase in depth sees little temperature change. The change in slope of this curve occurs, very approximately, at a depth of about 120 km and defines the base of the lithosphere and the top of the asthenosphere.

The red line defines the conditions under which the mantle peridotites can begin to melt. At temperatures and depths above and to the right of the red line, some melting would occur. In order for this to happen, and thus for basaltic magma to be generated, the black line must cross the red line and bring the mantle into the melting region. Thus, the left-hand diagram with no such intersection, represents the normal situation (e.g. beneath Scotland today) in which no magma is produced in the underlying mantle. By contrast, diagrams **A**, **B** and **C**, to the right show the three ways in which magma generation can be instigated and give rise to a potentially volcanic situation.

A represents the case where the mantle becomes hotter, e.g. as a result of a hot upwelling 'mantle plume', and the 'knee-bend' in the (black) temperature-depth curve moves right to cut across the (red) 'beginning of melting curve'. **B** shows the case where the lithosphere has been thinned by being stretched and the 'knee' in the temperature-depth curve now rises to shallower levels and, in so doing, intersects the 'beginning of melting curve'. Here, no heat has been added but the pressure has been reduced by thinning the lithospheric 'lid'. **C** represents the situation encountered above a subducted slab where, as a result of water entering the mantle peridotites, the 'beginning of melting curve' is modified into a bowed shape and the temperatures at which the 'wet' peridotite starts to melt are lowered. Here the red curve has been moved leftwards and cuts across the knee in the temperature/depth curve. In each of these three cases the deep-yellow shading defining the degree of overlap between the two curves marks the areas in which partly melted peridotite exists. The more the overlap, the more magma is produced.

behind subduction zones. These supra-subduction volcanoes, commonly spaced at intervals of 20–50 km apart, are strung around the Pacific like strings of pearls (*Fig. 1.10*). In various parts of Scotland we can recognise volcanic rocks that were formed in such supra-subduction situations.

When the loss of oceanic lithosphere by subduction exceeds growth along constructional plate boundaries, the ocean shrinks, ultimately vanishing as the converging continental masses on either side are brought into collision. One side may start to under-ride the other and the consequent crumpling and deformation of the rocks leads to the formation of a mountain chain. Such a mountain-building event is called an 'orogeny' and the whole process is referred to as 'orogenesis'. Modern instances most spectacularly include the Himalayas, where the Indian continent has moved north and crashed into, and partially under-ridden, the Asian continental mass to the north. Such mountain belts represent a more messy and more diffuse form of plate boundary. The continental rocks do not get recycled back into the mantle like their oceanic counterparts because they are much less dense and altogether too buoyant. Volcanoes are not abundantly represented along these continent–continent collision zones but nevertheless do occur. There are some, for example, along the great chain of the Alps–Carpathians–Caucasus–Elbruz–Himalayan mountains. Mt Ararat, of biblical fame, on the Turkish–Iranian border, is one such. A few even occur in Tibet, where magmas have managed to ascend through essentially a double thickness of continental crust. Volcanic rocks are observable in Scotland, where careful geological detective work indicates that they have the chemical hallmarks of supra-subduction origin, some from ocean–continent, and others from continent–continent situations.

Thus far I have considered volcanism associated with plate boundaries, the bulk being related to constructive and destructive plate margins. Volcanism, however, can occur in parts of the world remote from plate boundaries and is distinguished as 'intra-plate volcanism'. Intra-plate volcanism can be brought about simply by reduction of pressure (through extensional stresses) on the underlying hot mantle (*Fig. 1.9*) but it is facilitated if additional heat is added through the arrival of a hot mantle plume. Making the distinction between pressure-release magmatism without added heat (*Fig. 1.9B*) and that involving additional thermal energy (*Fig. 1.9A*) can be difficult and controversial. Irrespective of this controversy there are some good examples in Scotland of former volcanoes developed in intra-plate environments.

The Earth, unlike its neighbours, Mercury, Venus, the Moon and Mars, which are now inert or 'dead' planets, is an active planet with dynamic

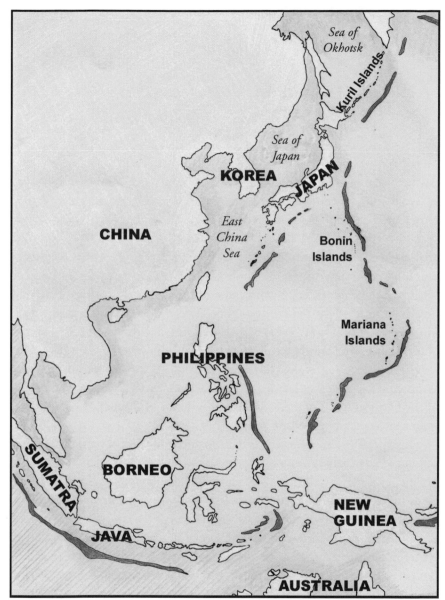

Fig. 1.10 A map of the western Pacific Ocean showing a pattern of volcanic arcs developed above subducting oceanic lithosphere. The dark zones are the oceanic trenches marking the site where the lithosphere commences its descent.

processes occurring in its interior. Although for mankind the results can be horrendous, earthquakes and volcanic eruptions may, nonetheless, be considered as symptomatic of a vigorous and living planet.

Chapter 2

Time on Earth and a Brief History of Scotland

As the hymn says, 'A thousand ages in Thy sight are like an evening gone.' It is difficult, to say the least, for us to appreciate the vast gulfs of time represented in the geological record. Historians of recent times deal in years whereas medieval historians may reckon in decades or centuries. Archaeologists and anthropologists concerned with the early history of mankind are happy to deal in thousands of years. It was James Hutton, born in Edinburgh in 1726, who came to the heretical conclusion that the Earth was much older than the 6,000 or so years claimed by the theologists. Hutton's great opus *The Theory of the Earth* ended with the immortal words: 'The result of our present enquiry is that we find no vestige of a beginning – no prospect of an end.' Hutton died in 1799 and in the ensuing two hundred years geologists learned to become cavalier in discussing the history of the Earth in terms of millions, tens of millions, hundreds of millions and even thousands of millions of years.

The planet is now believed to have formed from solid particles coalescing in the solar nebula around four thousand six hundred million years ago. Following a major collision with another planetary body resulting in the birth of the Moon, the Earth probably became wholly molten. Subsequent sorting out, or differentiation, of its components led to the generation of the metallic core and silicate-rich mantle. From the latter the crust developed, with the continental crust arising from processes specifically associated with the onset of subduction. For around the first six hundred million years of its existence our planet was subject to a merciless bombardment by space debris (asteroids, meteorites and comets) resulting in a highly pock-marked, cratered surface similar to those now preserved on the Moon and Mercury. Although our knowledge of the early developmental stages is entirely hypothetical since no recognisable materials are preserved from these remote times, our inert neighbouring planets provide irrefutable evidence that the Earth also experienced these cataclysmic bombardments. The arrival of extra-terrestrial materials continues to the current time but thankfully on a much reduced scale.

The oldest surviving terrestrial rocks, dated from their content of radioactive elements and their 'daughter' products, have ages approaching 4,000 million years. These occur in some continental regions such as, for example, in Australia, Greenland and the USA and include recognisable water-lain sedimentary rocks. The latter tell us that there was water on the Earth's surface and it is likely that familiar processes involving rain, rivers, erosion and sediment deposition were already operating in those remote times. There is also incontrovertible evidence that volcanoes were actively erupting in much the same way as volcanoes do today.

The oldest rocks in Scotland are those exposed in the far north-west, composing the Outer Hebrides and a narrow tract of ground extending south from the region of Cape Wrath to the Sleat Peninsula of Skye and to the islands of Tiree and Iona. These rocks, collectively known as the Lewisian from their widespread occurence on the Isle of Lewis, yield ages from 2,900 to 3,100 million years. The Lewisian consists largely of coarsely crystalline, striped grey and white or black and white rocks, referred to as gneisses, which are the end-products of long and complex histories. The ingredients that initially went into the making of these gneisses probably included sands and muds deposited in ancient seas over 3,000 million years ago. Lavas from volcanoes as well as intrusive igneous rocks almost certainly featured among the starting materials. Through repeated folding and recrystallisation during mountain building processes these original sediments and igneous rocks were so changed (or metamorphosed) into the gneisses that only the broadest speculations can be made as to their primary nature. Generally the radiometric ages (i.e. those given by the decay products of radioactive elements) reflect only the youngest of these metamorphic events.

By 2,400 to 2,000 million years ago the record becomes a little clearer. Dykes, which are essentially steep, parallel-sided, tabular bodies of igneous rock formed by the intrusion of magma (*Fig. 1.4*), are seen abundantly in north-west Scotland transecting the Lewisian gneisses. Dykes start off vertical or steeply dipping but can be folded or rotated to other attitudes by later earth movements. Before about 1,000 million years ago we may suspect that there were episodes involving the growth of continents and the rifting apart of continents once formed with the creation of new oceans. These oceans waxed (through growth along constructional plate boundaries), then waned (when rates of destruction along subduction zones exceeded those of growth). The resultant closure of the oceans, involving collision of the adjacent continents, caused uplift, folding and metamorphism, i.e. orogenesis. One such transient ocean may have opened and closed in the interval 1,100–1,000 million years ago, starting with the break-up of the great continent of Palaeopangaea, closing with the Grenvillian Orogeny and the formation of a continent called Rodinia.

From about 1,000 million years ago the geological record starts to be distinctly more complete (although large parts of the record are still missing!). Great thicknesses of sediment were deposited on the floors of shallow seas and/or from great river deltas debouching into the seas. Much of the landscape of the Grampians and the Northern Highlands has been carved out of rocks composed of these ancient sediments. At approximately 870 Myr an episode of lithospheric stretching caused a major outbreak of basalt volcanism. The extrusive products are not identifiable but great numbers of minor intrusions north of the Great Glen Fault probably represent intense swarms of dykes which would almost certainly have had surface expression as fissure volcanoes, either erupting on dry land (sub-aerially) or on sea floors.

Some time around 600 million years ago the over-stretched continental crust of Rodinia ruptured, allowing the generation of another ocean, Iapetus. Since the severed continental regions on either side largely correspond to those now seen in North America on the one hand and Europe and North Africa on the other, Iapetus is sometimes referred to as 'the proto-Atlantic Ocean'. The remains of lavas and intrusions generated through the volcanism attending the birth of this ocean can be seen at scattered intervals within the Grampians, but are seen at their best in the Tayvallich Peninsula on the west coast. Probably no sooner had the Iapetus Ocean reached its maximum width (whose magnitude cannot be assessed with any certainty) than subduction of the ocean floor below the continental crust on either side led to its shrinkage and ultimate closure. The latter involved the collision of the bordering continents, contorting and uplifting the rocks involved into a majestic mountain range. The subduction processes responsible for the demise of Iapetus were also the cause of extensive volcanism.

Since this mountain building episode, now known to have reached its peak at about 430 million years ago, was first recognised in Scotland during the 19th century, it is universally called the Caledonian Orogeny. The folded and faulted rocks marking the roots of this now deeply eroded mountain belt are recognised well beyond Scotland and compose much of the Appalachian Mountains in eastern North America, the Norwegian mountains and those of north-eastern Greenland. Although these pieces of the jig-saw puzzle are now widely separated through the movement of tectonic plates, they were once conjoined to form a great fold-mountain belt several thousand kilometres long and some hundreds of kilometres wide. Since the Caledonian Orogeny the only major tectonic crisis to affect Scotland occurred at approximately 60 million years ago, when the North Atlantic Ocean started its existence. This, and the volcanism attending it, are discussed in Chapter 5.

We can consider Scotland as a collage of continental slices (terranes) that were brought into juxtaposition by plate tectonics. We do not know exactly where on Earth these component terranes lay in earlier times but the study of rock magnetism provides useful clues. The Earth's magnetic field becomes locked, or fossilised, into the rocks as they form. Consequently, measurement of the rock's magnetism allows investigators to determine not only in which direction the Earth's magnetic poles lay at the time of the rock's formation, but also the angle of dip (declination) of the magnetic field. Since the declination changes with latitude, palaeomagnetic research can identify the latitude in which rock generation occurred. Such studies suggest that, by around 1,200 million years ago, the terranes that ultimately formed Scotland lay deep in the southern hemisphere. Certainly for the past 500 million years there has been a generalised northward migration. This migration brought them across the Equator some 200 million years later and on to their present day latitudes between 54° and 61°N.

Of the different terranes becoming recognised as composing Scotland, the principal five shown in *Fig. 2.1* are:

1) The Southern Uplands terrane, bounded to the south by the so-called Iapetus suture through northern England and to the north by the Southern Upland Fault;

2) The Midland Valley, lying between the Southern Upland Fault and the Highland Boundary Fault;

3) The Grampian Highland terrane between the Highland Boundary Fault and the Great Glen Fault;

4) The Northern Highland terrane, bounded on the SE by the Great Glen Fault and to the NW by a fault with a shallow easterly dip, called the Moine thrust; and

5) A terrane to the west of the Moine thrust, embracing the Outer Hebrides and a narrow strip of land along the NW margin of the Scottish mainland.

We know in great, and often horrific, detail the history of Europe through the 20th and 19th centuries, whereas much medieval history

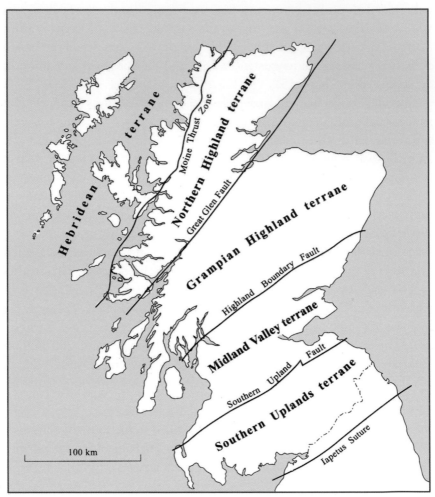

Fig. 2.1 Map of Scotland showing major structural units. *(After Trewin and Rollin, 2002)*

remains obscure. By the time we consider pre-Roman history the record is so sparse that only outline glimpses can be caught. Likewise our knowledge of the Earth's history becomes increasingly sporadic and inadequate the further back in time we go. The clear records are like oases in a desert; the more we attempt to unravel the events of the remoter past the fewer the oases and the greater the expanses of intervening deserts of ignorance and speculation. Nonetheless, occasionally in these metaphorical oases, we are treated to astoundingly vivid insights into the minutiae of the contemporary landscape and the prevailing conditions.

Instead of subdividing the last 4,600 million years into units of equal length, e.g. units 100 million years long, geologists have found it more convenient to subdivide it on the basis of significant events. Rather than keep repeating phrases like 'so many million years back', I will give numbers followed by 'Myr', standing for 'million years' and, unless otherwise specified, these will refer to that number of years ago. Distinct fossils made their first appearance around 540 Myr, when marine organisms developed shells or carapaces that had a reasonable chance of preservation. Major changes in the evolution of life, particularly as indicated by animal fossils, provide useful time breaks. There are three principal time divisions ('Eras'), primarily based on such palaeontological evidence. These are:

1) The Palaeozoic (derived from the Greek for 'old life'), extending roughly from 540 to 250 Myr;

2) The Mesozoic ('middle life'), *c.*250 to around 60 Myr; and

3) The Cainozoic ('new life'), extending from the end of the Mesozoic to the present.

Each of these is then sub-divided into a number of 'Periods'. However, in what constitutes modern Scotland we find no evidence of volcanicity in the Mesozoic record and, in the Cainozoic, volcanic activity was almost totally confined to the early part. It is when we start to look at the Palaeozoic that we find a long and complex history of volcanic eruptions. Accordingly it is useful to bear in mind the six periods into which the Palaeozoic is subdivided. In decreasing order of age these are the Cambrian, Ordovician and Silurian (composing the 'Lower Palaeozoic') and the Devonian, Carboniferous and Permian periods, which make up the 'Upper Palaeozoic'. A reference time-scale is presented in *Fig. 2.2.*

A huge period (approximately 4,000 million years long) that separated the creation of the planet from the appearance of such complex and 'hard shelled' marine creatures as trilobites, at the start of the Palaeozoic era, is simply referred to as the Precambrian. Much of the Scottish landscape is carved out of Precambrian rocks dating back, as noted above, to around 3,000 Myr. There is plenty of evidence within these pointing to repeated volcanism, although the detail is often obscure and wide open to speculation.

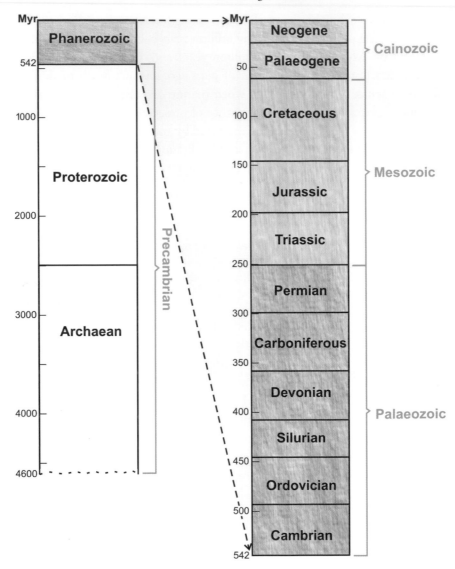

Fig. 2.2 The Geological time-scale, showing the approximate divisions between the periods. Phanerozoic ('visible life') is a term encompassing all of the Palaeozoic, Mesozoic and Cainozoic. Based on data from the International Commission for Stratigraphy.

Chapter 3

Magmas, Igneous Rocks and Volcanic Products

For those readers unfamiliar with the sciences of igneous petrology and geochemistry, this chapter is offered as an introduction. Since an igneous rock is defined as one formed by the solidification of molten magma, volcanoes are, therefore, by their nature and definition, largely or wholly made up of igneous rocks. Magmas are chemically complex but in the vast majority the elements oxygen and silicon predominate. Typically magmas cool to form rocks composed of crystals. Normally there will be several different crystal (or mineral) species present, almost always more than two but usually less than a dozen. Silicate minerals (i.e. those made up mainly of the elements oxygen and silicon) usually predominate among these. Solidification occurs either when the magma is cooled or when pressure is relaxed sufficiently for dissolved water, carbon dioxide (and other potentially volatile components) to come out of solution and escape to the atmosphere. Very commonly both factors operate in conjunction. Whereas, for example, molten wax or chocolate will solidify to a homogeneous material, this is not normally the case with magmas, whose solidification generally results in a heterogeneous rock made up of several mineral varieties that crystallised from the molten state.

Although discernment of the crystalline nature of igneous rocks may require the use of a hand-lens or microscope, the crystals are commonly large enough to see with the naked eye. Coarsely crystalline igneous rocks are often attractive when cut and polished. The facing stones on most city high-streets and bar-counters provide opportunities to examine the crystalline fabrics. Coarse granites, for example, can readily be seen to comprise a number of different crystal types (*Fig. 3.1*).

The dominant component will be feldspar, a term that covers a family of aluminium-bearing silicates which constitutes the most abundant mineral group in the Earth's crust. One, or commonly two types of feldspar may be present: the commonest in granite is potassium-rich and is usually pink or reddish brown. It may be accompanied by a distinctly separate feldspar that

Fig.3.1 Cut and polished surface of a coarse granite. Coin 20mm diameter.

is richer in calcium and commonly whitish in colour. Feldspar, whether as one or two different kinds, generally makes up about two-thirds of the volume of a granite while most of the remainder is quartz. Quartz is usually colourless or translucent grey. Together with feldspar and quartz there will normally be a speckling of blackish crystals composing up to around 10% by volume of the rock: these, containing most of the iron that is present, will be members of either, or both, the amphibole and the mica groups of minerals. As the crystals grow from the melt they will do so with geometrically defined, planar faces. In most coarsely-crystalline igneous rocks, however, it is impossible to see these since the growing crystals will, sooner or later, have impinged upon their neighbours and, in the closing phases of crystallisation as the last vestiges of melt disappear, the crystals will have interlocked, gaining irregular mutual boundaries.

The brief description above of a granite is given merely to exemplify the fact that igneous rocks are typically crystalline and usually involve several different crystal species. Fast solidification, as when the magma loses heat rapidly and/or when it loses its volatile components, results in great numbers of crystals being initiated in small volumes of magma. Consequently mutual interference between growing crystals occurs at an early stage and the resultant rock is fine-grained. Conversely, coarse-grained igneous rocks result either where cooling rates are very slow (as in great volumes of magma at depth) or where the volatile materials are retained under pressure. To revert to the champagne simile, the magma remains 'well corked' and keeps its potentially vaporous components in solution. Slow cooling and volatile retention usually go hand-in-hand.

In coarsely crystalline igneous rocks the individual crystals may be several millimetres to a few centimetres across. Indeed occasionally they may be large enough to trip over! Rock-climbers, for example in the Cuillin of Skye, gain extra grip from the differential weathering of the large crystals in the slowly-cooled rocks. Such coarse crystallisation normally takes place in large magma bodies ('magma chambers'), where 'large' usually means having length, breadth and/or depth measurable in hundreds or thousands of metres.

Smaller magma bodies intruded into the crust may adopt many forms. An extremely common form is as dykes, vertical or sub-vertical sheets with roughly parallel sides, as shown diagrammatically in *Fig. 1.4*. When revealed on the surface after erosion they frequently form upstanding walls since they are often composed of rocks tougher than those into which they are intruded ('country-rocks'), and the term 'dyke' comes from their frequent resemblance to stone walls. An example is shown in *Fig. 3.2*. Dykes are rarely more than a few tens of metres broad. Since dykes are commonly intruded in multiples, sometimes radiating from a focus but more often in sub-parallel bunches, they commonly consist of 'dyke swarms'.

Magma intrusions, particularly those reaching shallow levels (i.e. no more than a few kilometres from the surface), and encountering well strati-fied and low density rocks (generally sedimentary rocks), take advantage of the planes of weakness to spread laterally forming 'sills'. Sills are frequently approximately horizontal sheets, and it follows that their contacts tend to be parallel with those of their stratified (or bedded) country-rocks. As with dykes the thickness of sills is usually a matter of metres, although in excep-tionally large ones the thickness can exceed 100 m. In old volcanic terranes it can be difficult to determine whether an igneous sheet was an intrusive sub-surface sill or an extrusive lava flow.

Many sheet-like intrusions, however, conform to neither the dyke nor sill definitions but may be inclined at an angle and sometimes, as will be seen, have a crudely cone-like form. In some cases sheet-like intrusions have geometries defying simple description while other intrusions may have irregular, non-parallel margins and thus not be sheet-like at all. Magma filling sub-cylindrical or (usually downward-) tapering conduits cools to form what are variously described as plugs or necks. Often these are interpreted as filling the conduits that supplied magma to an overly-ing central-type volcano: the conduits may have been opened initially by

Fig. 3.2 A resistant dyke in NW Mull cutting softer basalt lavas that have been preferentially weathered.

violently escaping volcanic gases, sometimes reamed out by the subsequent ascent of magma.

Because there is a very rough relationship between the size of a magma body and the rate at which it cools, there is a corresponding relationship to the size of crystals in the rock formed from it. Even in the hundreds of small dykes, sills, plugs etc. to be encountered in Scotland, the crystalline nature is still apparent on close inspection. When cooling has been fast (accomplished within days, for instance) the products will still be crystalline but on a micro scale, so that a hand-lens or a microscope is required to distinguish them. However, when magma loses heat very rapidly, for example when erupted into water or dispersed as droplets into air during an eruption, there may be insufficient time for the constituent atoms to become geometrically organised into crystals. The magma thus fails to crystallise and instead congeals as a supercooled liquid into a glass. Rapid depressurisation with concomitant degassing can bring about the same effect. The shiny black natural glass known as obsidian is a product of fast cooling and/or degassing of a magma which, held in bulk and under pressure at depth, could have yielded a coarsely crystalline granite.

Failure to crystallise is particularly common in magmas rich in silicon, usually expressed in terms of its oxide, silica (SiO_2). These have greater viscosity than silica-poor ones and the viscosity impedes the migration of the atoms (or ions) to the extent of making crystal growth impossible. *Fig. 3.3* shows a dyke on Eigg where a combination of heat loss and degassing of a silica-rich magma has led to its congealing to obsidian.

Basalt magmas crystallise to some half-a-dozen different mineral species, although it is only important to bear in mind three of them. Plagioclase (a variety of the larger family of feldspars) is dominant: while the colour is variable, plagioclase is usually white or pale grey and, since it breaks easily along two planes controlled by the crystal structure to give shiny surfaces, it is usually easy to see the reflections from these smooth cleavage surfaces. The second critical component is augite (a variety of the pyroxene mineral family). Augite is either coaly black or very deep bottle-green. It too has two structural planes of weakness along which it cleaves readily, again providing shiny surfaces. A third mineral, which is common but not invariably present, is olivine. Olivine, in pristine condition, is translucent apple green and sold by jewellers under the name 'peridot'. Unfortunately olivine is rarely seen in such a condition in the Scottish rocks: olivine alters in the presence of water to secondary minerals that are more stable at the low-temperature and low-pressure conditions prevailing near the surface of the Earth. Commonly such hydration converts the olivine to a reddish or orange-brown material. Even when this has not occurred the olivine crystals generally contain so many minute impurities as to render them black and opaque and consequently unattractive for either lapidary or jewellery use.

The rocks generated by the crystallisation of basalt magma are referred to by the terms gabbros, dolerites or basalts according to their degree of coarseness. Gabbros are the coarsest, with the biggest crystals formed by slow, deep cooling whereas dolerite (average crystal size about 1 mm) is the preferred term for the rather faster cooled products with intermediate-sized crystals. Dolerite is a familiar rock type since it is a favourite material for kerb stones and the dark-coloured stone 'sets' of garden centres. It is also very widely used in crushed form for (dark grey) road metal. Basalt is the term employed for the still finer-grained rocks, typically as seen in lava flows and very small fast-cooled intrusions. To summarise, gabbro, dolerite and basalt all have essentially the same composition, are mineralogically alike and are all products of basalt magma; the essential difference is in their degree of coarseness (*Fig. 3.4*). Good examples of gabbros are provided by the rough-surfaced, coarsely crystalline rocks that form much of the black Cuillin of Skye, most of western Ardnamurchan and the hills

Fig. 3.3 Obsidian dyke on the island of Eigg, cutting basalt lavas. The dyke is approximately 1 m thick.

Fig. 3.4 A natural surface of gabbro cut by a swarm of small dykes. The bulk compositions of all the rocks are similar (basaltic). The gabbro is part of a large intrusion. After it had cooled slowly it was intruded by the small basaltic dykes. The dark margins of these are extremely fine-grained (virtually glassy) and represent the very first portions of the melts that came into contact with the cold gabbro and lost their heat quickly. The dykes dilated as more magma was fed into their axial zones. These later increments, largely insulated from the cold gabbro by the chilled marginal zones, cooled somewhat more slowly to give coarser basalt. They are still rather too fine-grained to be termed 'dolerites'. (*Hammer 80 cm length*)

of Ben Buie and Corra Beinn on Mull. Dolerite is extremely abundant, composing the thousands of dykes and other minor intrusions of the Hebrides and Western Highlands. Fine-grained basalt is the dominant rock forming the lava plateaux of e.g. northern Skye and north-western Mull as well as Antrim in Northern Ireland.

The term peridotite was introduced in Chapter 1 for a type of rock, mainly composed of the mineral olivine, that is a dominant component of the mantle. During the early stages of the crystallisation of basalt magma olivine crystals are abundantly produced. Processes separating the olivine crystals from their parent magma can give rise to olivine concentrates which, on cooling, form peridotites. Consequently although peridotites are mainly confined to the mantle they can also occur, particularly in association with gabbros, in crustal environments. In the eroded volcanic structures in Scotland notable examples of the gabbro–peridotite association are found in Rum and Skye.

There is a vast spectrum of types of magma and a bewildering range of igneous rocks can be formed from them. Although over 1,500 names have been used in the past for different varieties of igneous rocks, a recent commission recommended pruning this to less than 300. Their study is the remit of the professional igneous petrologist and geochemist. Since, however, this account of the ancient volcanoes of Scotland is not a textbook of igneous petrology, the complexities will be ignored and I will confine my use to only around a dozen very generalised igneous rock names. For the parent magmas I shall, so far as possible, confine myself to writing of basaltic magmas (from which basalt, dolerite and gabbro crystallise), granitic magmas yielding rhyolite, microgranite and granite and andesitic magmas. Basaltic magmas are relatively poor in silica and rich in iron, magnesium and calcium. In contrast, rhyolitic/granitic magmas are richer in silicon with less iron, magnesium and calcium (but concomitantly larger contents of potassium and sodium). Andesitic magmas are compositionally intermediate between the first two. Their slow crystallisation gives rise to coarse-grained diorites (*Fig. 3.5*).

Most important of all are the basaltic magmas. These, as have already been discussed, are the universal products of the mantle in all parts of the world and at all times in the Earth's history. Oceans cover the greater part of the globe and are predominantly floored by basalt, and basalt also covers large areas of the continents. Consequently, a Martian visiting Earth for a grab-sample of its rocks would probably leave with a specimen of basalt.

Granitic magmas (which could equally well be called rhyolitic magmas, the name of their fine-grained products), however, are essentially confined to the continents and are, for the most part, the products of the melting of older crustal rocks. They are richer in silica (and consequently crystallise quartz, a form of silicon dioxide) as well as in the alkali metals sodium and potassium which, during crystallisation, are accommodated in alkali-rich feldspars (*Fig. 3.1*). The element aluminium is a further major component. The acronymic term 'salic' is a useful one. It is an adjective emphasising that silica (giving the 's') and aluminium (giving the 'al') are important constituents. The term can be used in contradistinction to 'mafic', another

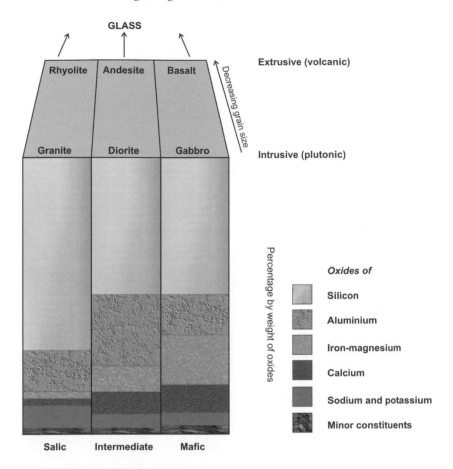

Fig. 3.5 Average chemical compositions and their relationship to common rock-names and grain-size. The acronyms salic and mafic are explained in the text. 'Plutonic' relates to coarse-grained rocks formed deep in the crust. *(After F. Press and R. Seiver, 1982)*

acronym built up from magnesium ('ma') and iron in its two oxidation states, ferrous and ferric, providing the 'f'. So salic embraces such compositions as granite, rhyolite and rocks (microgranites) texturally intermediate between them, while mafic covers such relatively magnesium- and iron-rich rocks as basalts, dolerites and gabbros *(Fig. 3.5)*.

The salic magmas have lower crystallisation temperatures (between 900° and 750° C) than basaltic (mafic) magmas (typically 1,160° to 950° C) and also have lower densities. Furthermore, salic magmas tend to have higher concentrations of water dissolved in them. When crystallised under pressure to yield coarse-grained rocks, some of the water enters the hydrous mineral species, amphiboles and micas, while much diffuses out into the

Property	Basaltic magma	Andesitic magma	Granitic magma
Silica content	Least (about 50%)	Intermediate (about 60%)	Most (about 70%)
Viscosity	Least *most fluid*	Intermediate	Greatest *most viscous*
Tendency to form lavas	Highest	Intermediate	Least
Tendency to form pyroclastics	Least	Intermediate	Greatest
Melting temperature	Highest	Intermediate	Lowest

Fig. 3.6 The relationship between common magma types, their silica (i.e. silicon dioxide) content and some of their physical characteristics.

Fig. 3.7 Ash-flow in Lascar volcano, Chile 1983. (*Jacques and Aïcha Guarino*s)

surrounding country-rocks. If, however, water-rich granitic magmas reach near-surface levels, the water escapes as vapour to the atmosphere. As it does so the viscosity of the magma rises dramatically and the struggle for water (and other 'volatiles') to escape from the stiff, stodgy and fast congealing siliceous magma, promotes violently explosive behaviour.

Whether we choose to call these silica-rich magmas rhyolitic or granitic is very much a matter of taste and in this account these two terms will be treated as synonymous. In many eruptions these magmas are not emitted as liquid lava flows but as particulate composite flows consisting of frag-mented magma (pumice) suspended in a gas matrix. Such gas–liquid suspensions ('ash-flows') travel as relatively dense aerosol-like materials at high velocities down-slope before collapsing and compacting as gas and energy are lost *(Fig. 3.7)*. Such ash-flows are capable of being channelled torrentially along river valleys and gorges. As they collapse, the hot, glassy pumice blobs can stick (or weld) together to produce a tough coherent rock called ignimbrite (*Fig. 3.8*). The welding may be so thorough that the ignimbrite may appear homogeneous, and expert investigation is needed to demonstrate that it was originally a particulate flow. *Fig 3.9* shows some of the ways in which these incandescent ash-flows may be initiated.

As granitic magmas approach within a kilometre or so of the surface portions or batches of the magma may exploit developing fractures in the surrounding country-rocks, feeding into them to form satellitic intrusions. Should the fracturing breach the surface the magma will commence violent degassing. The gases, dominated by water vapour, hitherto confined in

Fig 3.8 An ignimbrite (welded ash-flow product) on the island of Rum, Inner Hebrides. The lenticular fragments represent former glassy blobs of pumice that were flattened and deformed by compaction while still hot and ductile. Coin 25mm diameter.

Fig 3.9 Mechanisms of ash-flow eruption: a) collapse of a viscous lava dome allows gas-charged ash to avalanche down the volcano flank; b) a laterally directed explosive eruption from the lip of a crater; c) violent vertical discharge of gas and ash followed by d) gravitational collapse of the dense column producing ash-flows and atmospheric fall-out. *(After Myron G. Best, 1982)*

solution in the magma, will be blasted up and out into the atmosphere carrying globular masses of the magma with them. Gases in these globules will continue to come out of solution, forming a myriad expanding bubbles. As it loses its watery gas, while simultaneously cooling, the magma will solidify to very finely crystalline (or near-glassy) material, becoming a blob of pumice. Pieces of pumice are still fairly common objects in domestic bathrooms and will be familiar to most readers as pale-greyish coloured objects, light in weight because of the number of vesicles within them, making them rather like lumps of solidified froth. This sequence of events is depicted in *Fig. 3.10*, which shows the supposed evolution of a large magma chamber in the Mt Rainier National Park, Washington, NW USA.

The third principal magma type with which we need to concern ourselves is andesite. The andesitic magmas may be thought of as intermediate in chemical and physical character to the basaltic and granitic magmas (and are thus neither mafic nor salic: *Fig. 3.5*). The coarse-grained equivalent is

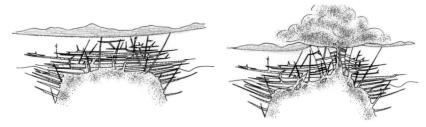

Fig 3.10 Mt Rainier: Tatoosh pluton. *Left*, fracturing above a large granitic magma chamber permits intrusion of small sheets into the overlying country-rocks. *Right*, magma eventually breaks surface giving rise to violent gas release and accompanying pyroclastic eruptions. *(After R. Fiske et al., 1963)*

known as diorite. Andesitic magmas tend to be particularly closely associated with subduction and are typical of those volcanic regions where ocean crust is subducted below ocean crust (the oceanic 'island arcs' such as the Marianas (*Fig. 1.10*) or the Aleutians) or where ocean crust descends beneath continental crust as, for example, in the Andes, from which they gain their name. Andesitic magmas tend to be sufficiently rich in water for explosive eruptions to be commonplace and the volcanoes are commonly 'composite cones' involving alternating strata of andesitic lavas and their pyroclastic products. *Fig. 3.11* shows a diagrammatic cross-section of an example in the Vanuatu arc in the SW Pacific. Mt Egmont (*Fig. 1.2*) is a good example of an andesitic supra-subduction volcano.

It is particularly such volcanoes that produce the most spectacularly violent eruptions with copious pyroclastic products. The eruption of Krakatoa in Indonesia in 1883 was the most celebrated one in the past one hundred and fifty years (although exceeded in energy release by another Indonesian cone, Tamboro, in 1815). In more recent history, eruptions of Mount St Helens in the NW USA (1980), El Chichon in Mexico (1982), Pinatobu in the Philippines (1991) and Popocatapetl in Mexico (2000) are all from volcanoes of this category which have hit the headlines. We shall encounter evidence for comparable Scottish supra-subduction volcanoes in the later chapters concerning activity between 390 and 490 Myr, from younger to older in the Devonian, Silurian and Ordovician Periods.

Unconsolidated pyroclastic material on the flanks of volcanoes gives rise to unstable deposits that can migrate down-slope by creep or as debris-flows. Heavy rainfall, as water condenses from eruption clouds, commonly accompanies big eruptions. The resulting water-sodden pyroclast deposits can move as mud-flows, sometimes with catastrophic consequences. A major tragedy occurred in Columbia in 1989, when the town of Amero was

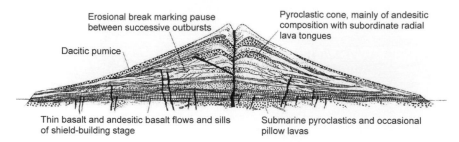

Erosional break marking pause
between successive outbursts

Pyroclastic cone, mainly of andesitic
composition with subordinate radial
lava tongues

Dacitic pumice

Thin basalt and andesitic basalt flows and sills
of shield-building stage

Submarine pyroclastics and occasional
pillow lavas

Fig 3.11 Section across Tongoa volcano, Vanuatu, SW Pacific Ocean. *(After A.J. Warden, 1967)*

overrun by mud-flows caused in this instance not by rain but by massive snow melting attending the eruption of Sierra del Ruiz. Volcanic mud-flows are commonly intercalated with lavas and other pyroclastic deposits.

As explained above, of the dozens of different names employed by specialists to describe the igneous rocks in Scotland, only the barest minimum will be used in this book, albeit at the cost of a considerable degree of precision. While bearing in mind these principal magma types of basaltic, granitic (rhyolitic) and andesitic composition, it should be remembered that not only is there a continuum between them but there are also some whose compositions cannot be encompassed within the basaltic–andesitic–rhyolitic spectrum. Thus there are some called trachytes which, like rhyolite, are formed and crystallised at relatively low temperatures and, again like rhyolites, have densities low enough to make them buoyant within the crust. Trachytes differ from rhyolites in having a higher proportion of alkali metals (sodium and potassium) to silicon, which results in their having higher proportions of alkali feldspar to quartz. Indeed, in some of the more extreme alkali-rich varieties quartz may be wholly absent. A further similarity between rhyolite and trachyte magmas is that they commonly contain high amounts of dissolved water and carbon dioxide, which are released as gas during the fall in pressure which occurs as the magmas approach the surface. Consequently trachytic eruptions are often explosive. Deprived of their volatile components any de-gassed trachyte lavas tend to be extremely viscous, with correspondingly restricted flow. The term salic – introduced earlier with respect to rhyolitic and granitic materials – is also appropriate to embrace trachytic magmas and their rock products. Trachytes in Scotland are essentially confined to the volcanic fields of south and central Scotland produced in the Carboniferous Period, between 350 and 320 Myr, and will be discussed in Chapter 6.

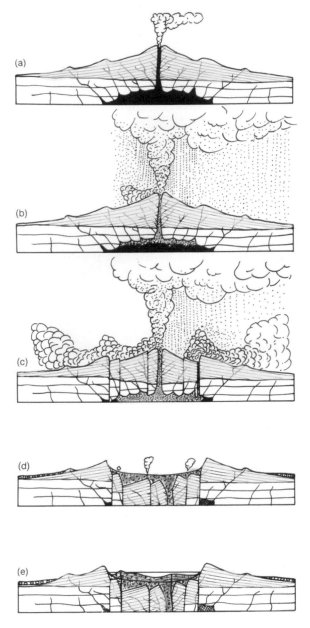

Fig. 3.12 Diagrammatic evolution of Crater Lake, Oregon, as a result of the big eruption some 7,000 years ago. (a) Prior to eruption; (b) some ash-flow and vulcanian ash-cloud at the commencement of the eruption; (c) the climax of the eruption, with significant ash-flows issuing from the central vent and ring fractures lower down the slope as the summit starts to sink; (d) settled into its current state, with newer eruptions on the floor and the caldera semi-filled with water. *(After H. Williams, 1942)*

How do all these different magma types originate?

Magmas are complex solutions: rather than solidifying by crystallisation at a simple freezing point they typically crystallise over a temperature range, commonly greater than 100° C. First one mineral species appears, to be joined by a second and, as crystallisation approaches completion, an assemblage of several different species crystallises together to produce the final rock. As an illustration, a cooling basalt magma will see olivine as the first major crystal species to appear. This may be joined by plagioclase or augite as a second species. By the stage of complete solidification (usually around 950° C) some five minerals (olivine, plagioclase, augite, magnetite and apatite) may constitute the end product. During the crystallisation process the chemical composition of the liquid changes progressively, since particular elements are abstracted from it by the growing crystals. For example, magnesium is taken from the liquid by the growing olivine crystals so that a magma that commences rich in magnesium may, in the last stages, be very poor in this element. Very commonly the silicon content rises as crystallisation proceeds and the amount of liquid remaining diminishes towards zero. The process bringing about these changes in the chemistry of the liquid is described as fractional crystallisation: it was recognised over eighty years ago that fractional crystallisation is an extremely efficacious agency for producing variation in magma types.

If, following the argument above, crystallisation of a basalt magma has proceeded to an advanced degree, the residual liquid (i.e. the last dregs before all of the melt has solidified) may differ strikingly from the original liquid in being depleted not only in magnesium but also in iron and calcium. It may be correspondingly enriched in silicon together with sodium and potassium, elements that were only modestly, or insignificantly, used up in the growth of the early mineral species. The composition in fact evolves to that of a rhyolite melt and the process involves continuous evolution from a mafic parent magma to a salic end product. If during natural processes such residual melts become separated from the already formed minerals they may solidify quickly to a glassy or fine-grained rhyolitic rock or slowly to a granite.

However, by no means all rhyolitic (granitic) magmas are created by fractional crystallisation from basalt magma parents. Much of the continental crust is composed of ancient rocks whose composition approximates to granite. These are readily raised to their melting temperatures, most commonly by heat given off from basaltic magmas arising from the deeper mantle. While rhyolitic/granitic rocks are present both under the oceans and in the continents they are overwhelmingly predominant in the continents, and the inference (confirmed by more sophisticated research) is that the continental granites are largely derived from remelting of older crustal materials. In brief, such silica-rich salic magmas have dual origins: some are wholly the product of fractional crystallisation while others are produced by the remelting of older rocks. Not surprisingly it has been established that many salic magmas had a hybrid origin involving both processes.

Since this book is not intended as a primer in igneous petrology, the origin of different sorts of magma will not be pursued further. Let it suffice to say that the extraordinary array of igneous rocks of differing composition can be mainly ascribed either to fractional crystallisation or to remelting of pre-existing, relatively fusible, rock types.

Chapter 4

Lava Flows and Pyroclastic Deposits

Before exploring the products of Scotland's old volcanoes it is expedient at this point to consider the nature of lava flows and the accumulation of fragmental volcanic rocks. As already stated, basalts are by far the most abundant lava type erupted by volcanoes, world-wide and from early in the Earth's history. Typically they are dark-grey to black, fine-grained rocks (although they commonly contain larger crystals, formed at depth, and carried up in the magma). Basalt lavas usually also contain bubbles (vesicles) caused by gases, formerly held in solution in the molten basalt, coming out of solution at the low pressure experienced on eruption. These are usually only a few millimetres across but can occasionally be much larger. On first formation the vesicles will be spherical but may be deformed into ovoids, tubes or more complex forms if lava movement continues after the end of gas exsolution. Thus, on eruption the lavas may consist of liquid (molten silicate), solid crystals held in suspension and gases separating from them.

Although basalt magmas contain dissolved gases, the content is typically much less than in andesitic and, particularly, rhyolitic magmas. Because there is less gas to escape, basaltic eruptions tend to be relatively quiescent in comparison. Despite this they can be sprayed up as fountains (*Fig. 1.6*) as a result of rapid degassing of the magma at low pressure. The growth of bubbles dramatically lowers the overall density, allowing the fountains to rise so high.

Basalt magmas are sufficiently fluid for the gases to separate with relative ease; consequently their explosiveness is muted in comparison with that of other more sticky and viscous types of magma from which separation of gas is difficult. Gases and small magma particles ('ash') may ascend to high altitudes and then be widely dispersed as air-fall ash. Much, falling into marine environments, may be diluted by muds and sands falling out from suspension as sediment. Larger (molten) particles or droplets fall to coalesce as lavas that may spread widely away from the fissure. Thus at Laki Volcano, referred to in Chapter 1 (*Fig. 1.5*), the lavas spread laterally some 45 km. In such mobile basalt flows the surfaces are essentially horizontal as would be the case, for example, of spilt milk or paint. Some of the larger Cainozoic lava flows of east Greenland are estimated to have flowed for distances of well over 100 km.

The molten magma droplets falling from eruption fountains coalesce to form spatter cones and frequently merge to form a coherent lava flow. Cooler and less voluminous basalt lavas flow with open channels bounded by higher ridges or levees. Towards the advancing front the central stream may subdivide, like the distributaries of a river delta. Such flows advance slowly, with cooling black masses of lava tumbling down the front, momentarily exposing incandescent material in the hotter interior. Lava flows of this kind are typified by sharp, ragged and blocky surfaces and are referred to by the Hawaiian name '*aa*' flows (*Figs 4.1* and *4.2*).

In contrast, hotter, more voluminous and faster moving flows are similarly known by the Hawaiian term '*pahoehoe*'. In these flows the hot mobile part of the flow is generally concealed beneath a congealed lid or carapace which acts as an insulating layer conserving the heat of the flow interior. The latter, which may be flowing at rates of metres per second, is usually confined to a tubular channel that may have a diameter of metres to tens of metres across. These convey the fluid lava to the front where it characteristically breaks out in lobate buds. The buds (or '*pahoehoe* toes') develop ductile rinds around themselves and inflate until they rupture to produce yet another bud (or toe). Because of the insulating efficiency of the flow carapace, such flows can continue to advance for great distances. Such lavas can also increase their thickness by a process of pulsatory inflation as more and more new lava is fed in from the vent. Consequently the flows are commonly composite, built up of many lava increments. While *pahoehoe* flows are often only a few metres thick, repeated inflations may give rise to thicknesses of many tens and, indeed, in some Precambrian flows, hundreds of metres. The viscous drag of fast moving lava beneath a semi-plastic or ductile upper surface can generate fold structures in the latter that can appear like coils of rope. Hence the use of terms such as 'ropy' or 'corded' to characterise *pahoehoe* surfaces (*Fig. 4.3*). In general *pahoehoe* lavas tend to have much smoother, often undulatory, surfaces, contrasting with the rubbly and jagged surfaces of *aa* flows.

Although *pahoehoe* flows are very common in basaltic eruptions, they are rarely encountered in the more silica-rich andesitic lavas and not at all in rhyolitic lavas. Andesite lava flows commonly exhibit *aa*-type features. Rhyolite lavas are relatively uncommon because their eruptions more typically produce disrupted pyroclast formations on account of the higher gas contents. However, rhyolite may be expelled as a lava after the magma has been largely de-gassed: such sluggish, high viscosity lavas rarely reach more

Fig. 4.1 Advancing *aa* lava, Heckla.

Fig. 4.2 Aa lava advancing through vegetation, Etna, Sicily.

Fig. 4.3 Rope-like features on the surface of a *pahoehoe* basalt flow, Reunion Island, Indian Ocean.

than a kilometre from the eruptive vent. They generally exhibit very rough blocky surfaces and may be composed very largely of glass (obsidian).

Highly viscous lavas frequently occur as bulbous masses blocking vents and filling craters and are referred to as lava domes. 'Dacite' is a term that refers to lavas that have compositions intermediate between those of andesite and rhyolite. Lava domes are typically composed of materials such as rhyolite, dacite and trachyte (referred to earlier), and are commonly slowly extruded, rather in the manner of giant blobs of tooth-paste, in the aftermath of an explosive pyroclastic eruption.

Pyroclastic formations are major components in many volcanoes, particularly where the magmas are enriched in potentially volatile materials. This is especially true for supra-subduction magmas in which dissolved water concentrations can be high. However, in contrast to the lavas, pyroclastic formations, particularly if not consolidated like welded ash-flows, are extremely susceptible to erosion. As a consequence we have to bear in mind that large proportions of the erupted pyroclasts rapidly become redistributed in contemporary sedimentary formations.

Gas discharged at high 'muzzle-velocities' from volcanic conduits, carrying quantities of fragmented magma and solid debris, can rise tens of kilometres into the upper atmosphere as 'plinian eruption columns'. Sooner or later, as the upper part of the column cools and increases in bulk density, it will spread out as a 'mushroom cloud'.

Fig. 4.4 opposite A tall plinian column of ash-laden gas rising above Kliuchevsky volcano, Kamchatka, Siberia.

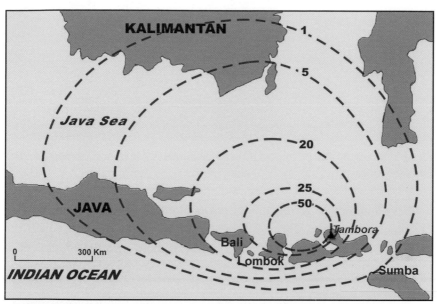

Fig. 4.5 Isopach map, Tambora Volcano, 1815. The dashed lines indicate the thickness of air-fall ash in centimetres. *(After P. Francis, Volcanoes; A Planetary Perspective, Clarendon Press, Oxford)*

Fig. 4.6 Bedded air-fall ash deposits near Hekla Volcano, Iceland. Dark layers, basaltic ashes; pale layers, rhyolitic ashes.

Fig. 4.7 Mount St Helens volcano, Cascade Mountains, Oregon, subsequent to the 1980 eruption.

Prevailing winds acting on eruption clouds will dictate the pattern of fallout from them. *Fig. 4.5* shows a typical fallout map in which the 'contours' (more properly, isopachs) indicate lines of equal thickness of fallout. Such air-fall will blanket the ground affected, the ashes draping over the landscape much as would a snow-fall. *Fig. 4.6* displays a section through some air-fall ashes, with changes of colour reflecting changes in magma composition during eruptions.

Unlike air-fall deposits, ash-flow materials tend to be channelled along valleys and any topographic lows. Although they will initially flow down-slope away from the eruptive centre, the high velocities of these hot, gas-lubricated avalanches may be sufficient to permit them to over-ride hills and ridges in their course. Their failure to drape evenly across the topography is one of the salient features distinguishing them from air-fall deposits. When the ratio of gas to particles is very high, the eruption is referred to as a 'surge', a term adopted from the atmospheric tests of nuclear weapons in the 1950s and 60s. Surges are laterally-directed blasts capable of attaining very high velocities. There are all gradations from ash-flow to surge deposits. Surge-type eruptions are frequently triggered where ground-water (in wet surface rocks, lakes, rivers etc.) becomes intimately involved with magma. Although pyroclastic deposits from surges tend to be rapidly eroded, some get preserved in the geological record. For example, in the volcanic vents that were produced in the tropical deltaic conditions of the Midland Valley in late Carboniferous times (Chapter 7), there is evidence of pyroclastic surges. *Fig. 4.7* shows the flank of Mount St Helens volcano (Oregon) that was opened up in a dramatic lateral pyroclastic surge in 1980. The partially dissected deposits from this eruption are seen in the foreground.

Chapter 5

Early Cainozoic volcanism and the Birth of the North Atlantic

The Cainozoic Era which comprises what were formerly the Quaternary and Tertiary Eras (in distinction to the Secondary and the Primary which are obsolete names for the Mesozoic and Palaeozoic Eras) began approximately 65 million years ago. It is divided into two Periods of which the earlier is the Palaeogene which ended at 23 Myr (*Fig. 2.2*). This was then followed by the Neogene which brings us up to the present time. Early in the Palaeogene volcanism broke out on a large scale in the Hebridean region. Although in this account I shall simply refer to all of this activity as Palaeogene it is referred to in many accounts, particularly the older ones, as being of Tertiary age. Consequently the rocks have very commonly been described as constituting the British Tertiary Volcanic Province..

Some sixty million years ago a major disturbance occurred deep in the mantle beneath the great landmass comprising what is now North America, Greenland and Europe, which culminated in continental break-up and the birth of the North Atlantic Ocean. A great upwelling of hot mantle rock (a mantle plume), is believed to have reached the base of the lithosphere and spread out laterally beneath it. Where there were thinner (and mechanically weaker) zones in this lithospheric 'lid', the lower pressures on the underlying mantle permitted some degree of melting of its component peridotites (cf *Fig. 1.9*). Simultaneously the lithosphere was being subjected to extensional forces and some of the newly generated basalt magmas were able to take advantage of the fractures and splits that were being created and arose as dykes.

The bulk of the magma, however, is inferred to have been trapped within the lithosphere around the level that separates the denser lithospheric mantle rocks from the lighter crustal rocks above. Of that proportion of the magma rising to form dykes in the crust, the great majority was fated to solidify before reaching the surface, as intrusive dyke rock. Only a minority of the dykes actually penetrated the full thickness of the lithosphere to erupt. Nonetheless, such was the magnitude of the convective upwelling in the deep mantle that thousands of cubic kilometres of basaltic magma were spilt upon the stretched continental surface in these Palaeogene eruptions. Remnants of these volcanic sequences, up to three kilometres thick, can now be seen around the coastal regions of north-east Canada (Baffin Island), west and east Greenland, the Faeroe Islands and in the British Isles. In the latter they occur in Northern Ireland and in the Hebrides from Arran to the north of Skye with some outcrops on the mainland in the Western Highlands. Great sequences, however, also exist beneath the offshore waters as, for example, in the Sea of the Hebrides and on the Norwegian continental shelf. The manifestations in the British region were, however, among the earliest of the mantle plume products, dating from approximately 60 Myr. *Fig. 5.1* is a sketch map showing the principal on- and off-shore outcrops of volcanic rocks in the north-western British Isles.

The volcanoes of East Greenland, Faeroes and the Norwegian shelf appear to have erupted on an even greater scale, some five million years later than those of Scotland and Ireland. These younger eruptions culminated when the extended and thinned continental litho-sphere finally parted along an ancient zone of weakness to give birth to the infant North Atlantic. From this time on the ocean has continued to grow, as a result of the new constructive plate boundary along the mid-Atlantic ridge taking North America and Greenland inexorably ever further from Europe. Whereas a massive uprise of hot (solid) mantle is inferred to have initiated the whole process, there is evidence that a much diminished stream or 'tail' continued to find its way up in post-Palaeogene times, persisting to the present day. The focus of this 'plume tail' is believed to underlie south-eastern Iceland and the whole North Atlantic mantle uprising is referred to as the 'Iceland plume', despite the fact that when it began, there was no Iceland, nor indeed had the North Atlantic Ocean started to open. All Icelandic volcanism is deduced to be fed, both in terms of energy and material, from this plume tail.

Fissure and shield volcanoes

It is the formations reflecting the onset of the Iceland plume event that give rise to so many of the most scenic features of the Inner Hebrides and parts of the west Scottish mainland. These include the lava plateaux of NW Mull, the Treshnish Islands, the 'Small Isles' of Muck, Canna and Eigg, northern Skye and parts of the Morvern and Ardnamurchan peninsulas on the mainland (*Figs. 5.2–4*). Individual lavas are typically less than

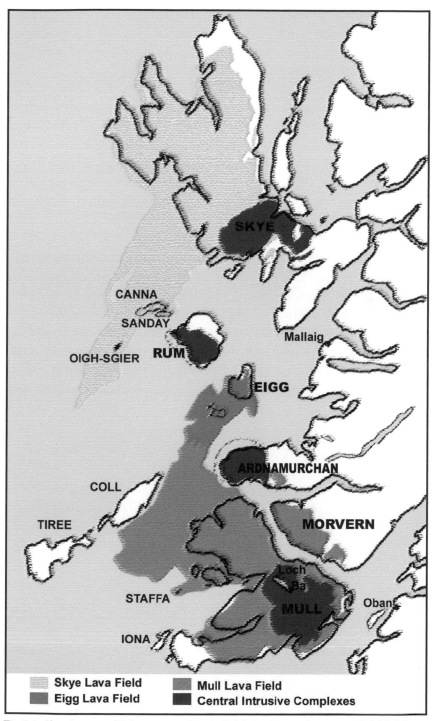

Fig 5.1 Sketch map of Palaeogene volcanic fields in the north-western British Isles. *(After Bell and Wilkinson, 2003)*

10 m thick although some, especially where they have been able to 'pond' in hollows or blocked valleys, are very much thicker. Both *pahoehoe* and *aa* type flows are present; 'more evolved' lavas, usually with higher silica contents than the basalts, are characteristically thicker.

These lava sequences now rarely exceed a kilometre in thickness and the original lava piles were probably not very much thicker than those surviving, for instance on Ben More, Mull or the high plateau of the Trotternish Peninsula of Skye. While it is inferred that the bulk of these lavas erupted from dyke-fed fissure volcanoes, doubtless some was also emitted from central-type shield volcanoes. The fissure volcanoes were supplied, as explained above, by magma rising through dyke conduits, commonly 1–5 m wide but occasionally very much bigger.

Extensional stresses generated across the northern part of the British Isles were such as to generate great numbers of dykes, the majority of which trend NW–SE. They are seen in their greatest abundance in the Hebrides and Western Highlands, although they also appear elsewhere in Scotland, England, Ireland and to a very minor extent, in Wales. The Palaeogene dykes of Scotland, however, are not evenly distributed but are grouped or concentrated into swarms, some tens of kilometres across. The most notable of these are the Skye swarm, the Mull swarm and the Arran swarm. Although the dykes occur in profusion over a large area,

Fig. 5.2, opposite Palaeogene lava scenery near the Quirang, northern Skye.

Fig. 5.3, top Palaeogene lavas on Horse Island, close to Muck, Inner Hebrides.

Fig. 5.4, above Palaeogene lavas on Canna, Invernesshire Small Isles.

Fig 5.5 The long promontory on the east side of Camas Mòr, Isle of Muck, is formed by an unusually broad member of the Hebridean dyke swarm.

some localities where they can be easily appreciated are the SE coast of Mull between Loch Buie and Craignure, the west coast of Eigg in Laig Bay and the south coast of Arran. These last differ in having a more nearly N–S orientation and form prominent rock walls rising from the more erodable sandstones that they penetrate. The very ragged, spiky map of Muck is the result of its many dykes forming promontories while the weaker lavas have been cut back into the bays. Muck boasts one of the larger examples of the Palaeogene dykes, well seen on the south coast, bounding the eastern side of Camas Mòr (*Fig. 5.5*). This intrusion, nearly 200 m wide, cooled sufficiently slowly for the rocks to be coarse enough to merit the term gabbro. Many dykes can only be traced for short distances but some of them can be traced for great distances: two or three dykes of the Mull swarm, for example, can be mapped south-east for several hundred kilometres across southern Scotland and northern England to the North Sea coast.

As already stated, most dyke magmas never reached the surface but lost heat *en route*, solidifying at depth as intrusions. How does one distinguish those dykes that did reach the surface and produce a fissure volcano? The answer must be 'With difficulty'. However, basalt magmas have temperatures of over 1,000° C and, in sufficient quantity, have the capacity to bake, or thermally metamorphose, the rocks through which they pass. This metamorphism generally entails a toughening of the rock as its component crystals grow and interlock as they heat up. The degree of metamorphism depends critically on the length of time that the country-rock is subjected to the heat of the magma. A baker who leaves his buns in the oven for five minutes gets a very different product from what he gets if he leaves them in for three hours, even if the temperature of the oven remains constant! Consequently a volume of magma injected as a dyke (a process that probably involves only hours or days at most) which 'dies of cold' as a subsurface intrusion has relatively little energy to transfer to its wall rocks and the baking effects will be minimal. By contrast, a 'feeder dyke' that supplies a surface eruption may act as a conduit for high-temperature magma over an extended period – perhaps months or even years – and the diffusion of thermal energy to its walls will be much greater. Rocks tend to be very good insulating materials, hence transfer of heat by conduction from the feeder dyke is a slow process.

Many rocks, especially poorly consolidated sedimentary rocks (shales, sandstones etc.), contain substantial quantities of interstitial water and it is commonly the case that much of the heat transfer is brought about by the migration of very high-temperature aqueous fluids. At temperatures well above normal boiling point and at comparatively high pressures water, instead of turning to steam, acquires the remarkable properties of a 'super-critical fluid', which is neither liquid nor gas but a separate state of matter. These super-critical fluids have extraordinary penetrative powers and can find their way through what might appear to be a totally impermeable rock, as well as having enhanced solvent properties. As a result much of the heat transfer between a magmatic intrusion and its wall rocks is achieved by convective processes involving diffusion of super-critical aqueous fluids. A dyke that acted as a magma conduit for a long time can often be recognised by the extent of metamorphism of its wall rocks. Again, Laig Bay on the island of Eigg provides beautiful examples where intensely baked (thermally metamorphosed) sandstones form upstanding walls alongside quite small dykes solidified in conduits through which huge volumes of molten basalt must have passed to feed fissure eruptions (*Fig. 5.6*).

Fig 5.6 Thermally metamorphosed sandstones adjacent to a small dyke in Laig Bay on the west coast, Isle of Eigg. Dykes like this may have fed fissure eruptions above.

Palaeogeography of the early Tertiary landscape

The landscape across the Northern Ireland and west of Scotland regions in the early Palaeogene, before the onset of volcanism, appears to have been one of relatively gentle topography. A number of subsiding basins which had been established during the preceding Mesozoic Era persisted into the Palaeogene (*Fig. 5.7*). These basins or troughs were defined by faults which sometimes bounded them on both sides or, more eccentrically, on one side only. The trend of the faults, the basins and the intervening strips of higher ground was dominantly NE–SW, following an ancient 'grain' in the basement rocks established several hundred million years earlier in Palaeozoic times. One prominent trough, encompassing most of what is now Skye, lay between the Outer Hebrides and the north-west Highlands. Another was to the south-east of a ridge of higher terrane embracing Coll, Tiree, Rum and part of southern Skye. Islay, Jura and Kintyre were also parts of elevated land further south, with NE–SW troughs extending across much of Antrim and another through the Arran area. Sedimentary strata, consisting mainly of shales, sandstones and limestones, had been deposited from shallow seas in these troughs during interludes of raised global sea levels during the Jurassic and Cretaceous periods in the Mesozoic. These strata had experienced uplift, gentle folding, faulting and subsequent partial erosion before the earliest eruptions began. To a very large extent it was these same Mesozoic troughs that continued to subside after the start of volcanism, thus accommodating the successive lavas erupted within them. Their subsidence tended to keep pace with the accumulation of the lavas so that the surface of the nearly flat-lying volcanic fields remained generally above, but close to, sea level. The troughs appear to have been largely occupied by shallow lagoons, lakes and swamps when the first magmas broke through to the surface in response to lithospheric stresses and heat supplied by the mantle plume.

The initial eruptions were generally explosive as a result of the generation of steam attending the interaction of the magmas with surface waters. Much of the basaltic magma was sprayed up as droplets which chilled to glass. These then splintered and fragmented through thermal contraction and, while still hot, absorbed water, producing hydrated basaltic glass. Subsequent cementation of the accumulated glassy fragments by migrating solutions produced a type of rock known as hyaloclastite. This word means 'broken glass rock', although crystallisation of the glass ('devitrification') would have followed swiftly, yielding a fine-grained assemblage of hydrated silicate minerals of the clay and chlorite families.

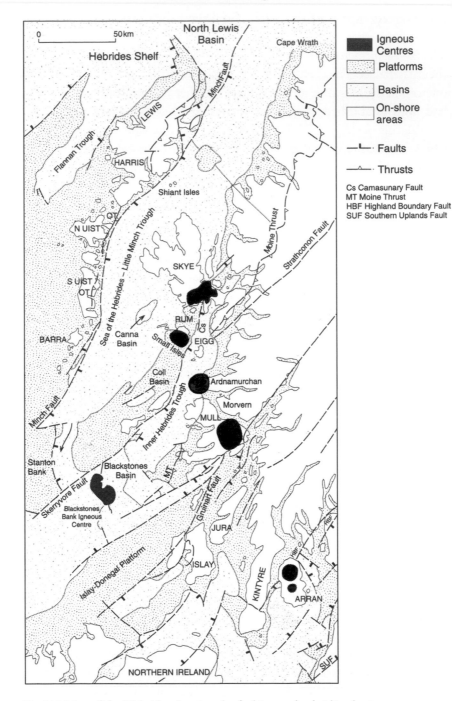

Fig 5.7 Map of the Hebrides showing the faulting and subsiding basins.
(After Bell and Wilkinson, 2003, The Geology of Scotland, 4ᵗʰ edition).

The interaction of hot magma and cold water, however, led to products other than hyaloclastites. Some of the fine particles blown up into the atmosphere in steam-driven explosions fell back into the waters, where they became diluted by muddy particles swept in by rivers. The resultant mixture of terrestrial detritus and volcanic ash produced volcanic mudstones and ashy shales. Elsewhere lavas which erupted below the waters tended to ball up and form lobate and rounded sack-like masses, referred to as pillow lavas. Occasionally magma was erupted into thick muddy sediment to produce elongate, sausage-like, 'mini-pillows' of basalt separated by a matrix of mud, or into mixtures of mud and glassy shards. As with the hyaloclastites, the latter quickly lost their glassy character during recrystallisation, degrading to a fine-grained assemblage of silicate crystals.

The early Palaeogene afforded opportunity for the rapid evolution of plant and animal species. We are dealing with the time during the first few million years after the global catastrophe or catastrophes which caused the extinction of many groups of organisms, that ended the Mesozoic Period. The dinosaurs, which had been the dominant vertebrates during the Mesozoic, were among the casualties. New forms of flora and fauna evolved rapidly to exploit the myriad environmental niches left vacant by the species that had failed to survive. From the frequent occurrence of charred (and usually silicified) fragments (e.g. twigs, leaves, rootlets) of plants among the Hebridean Palaeogene strata, we get a picture of a well vegetated landscape. Conclusive evidence for the lush vegetation in the lowlands comes from the thin coal-seams that are here and there interbedded with the volcanic products. While such seams are typically very thin (rarely more than a few millimetres) one needs to remember that the formation of even a thin coal-seam requires the compaction of a much greater thickness of accumulated plant debris. The inference is that in some of the subsiding troughs material from surrounding forests was being swept into the lakes and swamps by rains and streams over hundreds or thousands of years, to be progressively compressed by the mass of overlying detritus.

From all the above a picture emerges of a forested landscape with swelling hills bordering valleys and basins containing ample groundwater. This arcadian scene must have been rudely interrupted at the onset of volcanism by earth tremors, opening fissures and the spilling of incandescent lavas.

Much of the evidence for the statements above comes from the excellent geological sections around the Hebridean coasts. On Mull, for example, some of the ashy mudstones denoting the start of volcanism are visible beneath the lavas round the east coast, near Grasspoint. On the south coast of Mull, cliff sections and wave-cut platforms east of Carsaig provide dramatic exposures of mixed mud and lava, associated with thin coals and shales with plant fossils. Assemblages involving pillow lavas and hyaloclas-tites beneath the main lava sequence are well exposed on the coast of Skye in the vicinity of Portree.

Probably as a consequence of magma feeding laterally as sill intrusions into the Mesozoic strata beneath the Hebridean rifted troughs, there was relative uplift of the land and drainage of most of the swampy lagoons that existed at the onset of volcanism. While water-courses and lakes were still present, the geological evidence clearly shows that for most of the three million or so years of magmatism, the lavas and ashes accumulated on dry land – i.e. sub-aerially – in contrast to the sub-aqueous nature that charact-erised much of the earliest eruptive episodes.

From the various pieces of evidence a picture of the Hebridean region comes into focus. A pattern of lowland basins in what is now Antrim, parts of the Firth of Clyde and the Sea of the Hebrides, was bounded to the east and south-east by hill country in the region now occupied by the Grampian and Northern Highlands and by the Southern Uplands. The Outer Hebrides as well as Coll and Tiree represented ridges separating some of the lowland basins. Fluviatile deposits in the form of sand and pebble beds sandwiched within the lavas testify to a well-watered landscape with abundant rivers and streams (*Fig. 5.8*). However, with the Atlantic Ocean still in gestation, the rainfall is likely to have been much less than that currently 'enjoyed' by the Hebrides. Evidence from both coarse pebble beds (conglomerates) and indications of steep-sided valleys indicates a fairly vigorous topography with fast-flowing streams.

Some of the finer silts and muds found in the sedimentary inter-calations in the lava pile may represent over-bank flood deposits, while others probably accumulated in shallow lakes. The most celebrated of such palaeo-lake deposits are those underlying several metres of pebbly grits at Ardtun, near Bunessan in SW Mull. Beautifully preserved leaf-fossils occur in these but, because of extensive excavations, good material is now only recoverable with difficulty, and can be more conveniently examined in museum collections, for example those in Glasgow, Edinburgh and London. The leaves of some of the genera are very similar to those growing at the present time, e.g. oak, hazel, walnut, plane and ginkgo (maidenhair tree), indicative of mild and thoroughly agreeable climatic conditions. Prominent global warming, which commenced from about 59 Myr, was to reach a maximum by about 51 Myr.

Ongoing palaeo-botanical research into fossil leaves and, increasingly, into spores within ancient soil horizons trapped between lava flows allows

Above left, Fig. 5.8 River-floor pebble bed (conglomerate horizon) within the lava sequence, Isle of Canna. *Above right, Fig. 5.9* MacCulloch's tree, Mull. *(C.H. Emeleus)*

us insights into the vegetation that clothed the contemporary landscapes. Equisitales, ancestral to the horsetails familiar as a garden weed or as miniature swamp forests in Highland lochans, were abundant. Although growing to several metres in height, these were pygmies compared to their coal-forming ancestors in Carboniferous times, 230–240 million years earlier. Fragrant coniferous forests probably covered much of the landscape, with Metasequoia, a left-over from Jurassic times (and found still surviving in China in the 1940s), swamp cypresses and early species of pine.

When a basaltic flow a few metres thick encounters a forest the lava quickly cools against the tree trunks and the wood inside progressively chars and dehydrates. Those parts of the tree standing up above the flow surface heat up, dry out and ultimately burst into flame. The superstructure of the tree then collapses onto the flow surface and continues to burn until virtually nothing remains. Vestiges of tree trunks and branches occur within the lavas at several localities on Mull, including the famous 'MacCulloch's tree' on 'the Wilderness' (*Fig. 5.9*). This particular fossil is of an upstanding tree trunk some 12 m high and is believed to have been a type of swamp cypress.

Sporadic eruptions gave lavas that temporarily dammed the valleys allowing lakes to form: a more recent instance of such a lake is Lake Aydat in the Massif Central of France, formed some tens of thousands of years ago (*Fig. 5.10*). The Palaeogene landscape probably teemed with life: assorted fish, amphibians and reptiles (such as turtles and crocodiles) swam in the waters; the forests would have been home to snakes, lizards, birds and a multitude of beetles, snails and other invertebrates. Primates, the group of mammals to which man belongs, possibly evolved in Africa and spread to northern continents in the late Palaeogene. These small tarsier-like tree-dwellers may well have lived in the warm forests of the Hebridean region.

As a fissure eruption proceeds, parts of the dyke conduit through which the magma is fed may become choked by solidifying basalt. Further flow then becomes more and more concentrated into specific nodal points so that what may have started as a continuously active split gives way to a chain of separate cinder cones. The eruption may evolve further so that all supply becomes focused into a single pipe that is ovoid or even circular in cross section. In other words there would sometimes have been evolution from fissure-type to central-type volcanism. In terms of time, the actual eruptions would have varied widely. The fissure stages may have been accomplished in days or a few months but in other cases, particularly when a more nearly cylindrical feeder pipe had been established, eruptions may have persisted for several years. Magma flowing through these localised pipes would have supplied lavas to low-angled shield volcanoes, whose form was like that of a great up-turned saucer, with a central crater. The shield volcanoes would have appeared very much like the recent (i.e. post-glacial) shields in Iceland (*Fig. 5.11*). The feeder pipe itself would typically have had a diameter of up to 500 m. The height above the surroundings of the sort of lava shields fed through such pipes would rarely have exceeded 100 m, and the external diameter of the shield would probably have been less than 15 km. One good example of such a pipe occurs at S'Aird Bheinn in NW Mull.

It is important to bear in mind that the original surface expressions of these ancient volcanoes and the former appearances of the lava surfaces have almost always been lost through erosional processes. There has been very extensive erosion during the fifty or more million years or so since the eruptions came to a close, much of it accomplished by the glaciers of the

Fig 5.10 Lake Aydat in the Chaîne des Puys, Massif Centrale, France. An example of a lava-dammed lake in a forested region. Allowing for the difference in the vegetation, this scene may be a fair approximation of a Mull landscape in the Palaeogene.

Fig 5.11 Skjalbreidur volcano: a basaltic shield volcano in Iceland.

Ice Ages in the past few hundred thousand years. Furthermore, a lot would have taken place more or less straight away after eruption during the long periods of quiescence between one eruption and the next. These quiescent intervals were commonly of several hundred or even several thousand peaceful years when nothing very dramatic happened. While the rainy and dry seasons came and went, seasonality is likely to have been less marked than it is today.

In the warm and humid conditions of Palaeogene Scotland the surface features of the lava flows would have been ephemeral and rarely preserved. From modern analogy it is likely that algae, mosses and ferns were among the first plants to colonise following an eruption, growing amongst the nooks and crannies of the lava tops within a few months of their cooling. These would have formed the botanical advance guard, preceeding growth of conifers and broad-leaved trees as afforestation was established. Not only did the lavas provide sheltered anchorages for seedlings but they retained moisture and supplied the requisite fertilisers. Glassy material, the fastest cooled parts of the flow tops that never had time to crystallise properly, is relatively soluble and magnesium, calcium, sodium, potassium and a host of minor elements necessary for plant growth would have been available to the flora. Importantly, relative to many rock-types, basalts have relatively high contents of phosphate which, as every gardener knows, is vital for healthy plant growth. The plants themselves would have promoted the breakdown of the flow surfaces, partly by the mechanical hydraulic action of roots, which exploited and wedged open cracks, and partly by the secretion of humic acids. Thus, fertile soils, developed as a result of the chemical and mechanical breakdown of the lavas, augmented by the debris of decaying plant matter, developed on the forest floors. The original ropy or blocky nature of the lava surface was steadily erased and, by the time – perhaps centuries later – when the next lava eruption inundated the area, the volcanic terrain formed by the previous eruption had already been changed out of recognition.

Percolation of water through the decaying flow tops led to removal of the more soluble components and hence concentration of the more insoluble. Much of the magnesium, calcium, sodium and potassium, for example, was leached out ultimately to end up in the sea. This left the flow tops relatively rich in aluminium and iron. The iron became oxidised in the process forming its red oxide, haematite (so named because of the resemblance of its colour to that of blood!). Consequently the flow tops were converted from the original dark grey or black pristine basalt to an earthy, rusty-red or orange coloured product. Sometimes, at the top of these red weathered rocks ('boles') relics of soils are preserved, containing spores or fragments of the vegetation that grew upon them. The reddened boles are easily seen in cliff sections, river gullies and road-cuts (*Fig. 5.12*).

Lavas will naturally flow down-slope along whatever valleys the land-scape offers. However, streams and rivers, though their courses may have been interrupted, are never stopped. The rain still rains on the highlands, no matter what a volcano may be doing elsewhere. A stream blocked by a lava flow responds by flowing alongside it or even along both sides of it, carving out a new valley or valleys. In this manner what was the bottom of the original valley may evolve to become a ridge of relatively high ground

Fig 5.12 The reddened horizon represents the leached and oxidised surface of a basalt lava, overlain by a younger flow. NW Mull.

Fig 5.13, above The Sgùrr on the island of Eigg viewed from the east. (*C.H. Emeleus*)

with new valleys beside it. The process gives rise to topographic inversion: what was low ground becomes a positive topographic feature and vice versa. Sands and pebbles in a stream bed will be overridden and preserved by the next lava flow that follows the course of that stream. The steep edifices of Preshal More and Preshal Beg that dominate the Talisker Valley in north Skye would once have occupied a valley whose flanking hillsides are long since gone. The ridge of the Sgùrr that forms the very prominent eminence on Eigg is perhaps the most striking instance of such inverted topography in Scotland (*Fig. 5.13*). The Sgùrr is the erosional remnant of a very thick (over 200 m) flow that was at least partially channelled along a deep valley eroded into earlier basaltic lavas. In the case of the Sgùrr, we are not dealing with a basalt but a dacite whose composition more closely approaches that of a rhyolite. Beneath the flow, sands, gravels and boulders of the former river responsible for the palaeo-valley can be seen (most easily on its southern side). Fossilised conifer wood also occurs in these ancient river deposits as reminders of the forests through which the river flowed.

Inter-flow river deposits of pebbles (consolidated to the type of rock called conglomerate) can be seen in the sea-cliffs north of Glen Brittle, Skye, and are also prominent among the Canna lavas (*Fig. 5.8*). Detailed mapping in the west of Rum has revealed a story of river erosion and conglomerate deposition followed by lava flow repeated through several cycles among the hills of Orval, Bloodstone and Fionchra (*Fig. 5.14*).

Fig 5.14 The terraced features reflect several lava flows erupted in successive stream valleys, Orval, western Rum.

Fig 5.15 Aerial view along volcanic lineaments of the Reykjanes peninsula, SW Iceland in the vicinity of the Naesevellir geothermal area. Views along the fissured landscapes above the principal dyke swarms in the Hebrides would have appeared similarly in the Palaeogene.

The landscape above the main dyke swarms was probably one of sub-parallel topographic ridges and trenches with an overall NW–SE orientation. Pull-apart fracturing (or 'normal faulting') would have produced linear fault blocks, commonly tilted with a steep scarp on one side and a gentler dip slope on the other. Chains of cones and craters would have delineated major eruptive fissures while open fissures would probably also have played a role in defining these topographic lineaments (*Fig. 5.15*). Craters and fissures left in the immediate aftermath of eruptions would have filled with water, but erosional processes would generally have erased most of these features before any resumption of volcanism.

Lavas flowing over an eroded landscape would not only have been constrained by valley walls but would also have been impeded by previous flows or fault-scarps. In such cases ponding occurred with the obstructed lava attaining thicknesses of tens of metres. The cooling of the lava flows is invariably attended by thermal contraction that creates cracks or joints within them. Analogous contraction cracks, but due to water loss rather than cooling, are to be seen in any muddy puddle as it dries out, producing polygonal patterns. Lava flows will lose heat from their tops and bottoms and slow cooling of a thick lava allows the very regular development of such joints; the fractures propagate upwards from the floor as cooling proceeds, producing hexagonal columns that are extremely regular in terms of size and shape. Heat loss from the top is more rapid, often aided by rain-water or streams developed on top of the lava. The faster cooling in the upper part of the lava allows a more irregular cracking pattern to develop. These closer spaced cracks (joints) typically propagate downwards and ultimately meet the up-growing joint columns in a rather sharp boundary. These features may be clearly seen in old thick lavas, often best revealed in coastal cliffs. The terms 'colonnade' and 'entablature' have been borrowed from architecture to describe these features, with the former describing the lower elegantly columnar portion and the latter employed for the more chaotic upper parts. The most famous example in the British Isles is that of the colonnades exposed in Palaeogene basalts at the Giant's Causeway in Antrim. However, scarcely less celebrated are the beautiful columns exposed on the Island of Staffa in the Treshnish Isles, north of Iona (*Fig. 5.16*). At a distance the colonnade resembles a bundle of upright sticks, hence the name, which derives from the Norse, meaning 'isle of staves'. On Staffa both colonnade and entablature are readily seen. Fingal's Cave, which so attracted Mendelssohn's attention, was formed where wave action selectively plucked out some of the columns.

Although the Giant's Causeway and Staffa offer the most celebrated examples of columnar jointing in the early Palaeogene lavas, further excellent

Fig 5.16, above left Colonnade and entablature in a thick basalt lava, Staffa, Inner Hebrides. Staffa. *Fig 5.17, above right* Preshal More, a thick lava flow on Skye exhibiting colonnade and entablature features.

examples can be seen, e.g. near Ardtun in SW Mull and along the southern coast of much of the Ross of Mull as well as in lava flows on Skye, particularly those of Preshal More (*Fig. 5.17*) and Preshal Beg near Talisker Bay.

Smaller scale features apparent to anyone who gives more than a cursory glance at the lavas are 'amygdales'. These objects, ranging in size up to about 5 mm and from spherical to totally shapeless, may be scattered through a lava flow or concentrated in specific horizons, but are usually most abundant towards the top. They are called amygdales from a supposed resemblance to almonds (Latin '*amygdala*'). In the Hebridean Palaeogene lavas the amygdales are usually white and composed of small crystals that are frequently present as elongate needles or prisms. These crystals, mostly belonging to the mineral family of zeolites, were deposited in former gas bubbles by the permeation of hot (over 100° C) watery solutions through the basaltic rocks during, or (geologically speaking) soon after the cessation of volcanism (*Fig. 5.18*). The importance of such hydrothermal solutions, generally in a super-critical state, has been referred to earlier. These solutions react with the already formed igneous rocks, modifying their mineralogy, dissolving components in one place and depositing materials (e.g.

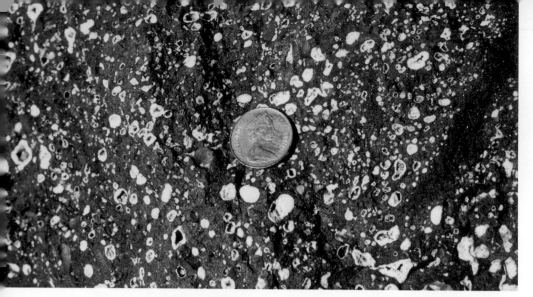

Fig. 5.18, above Zeolite-filled vesicles (amygdales) in basalt lava, Isle of Eigg..

zeolites) elsewhere. However, they would also have had surface expression, giving rise to hot springs, geysers and gas jets (fumaroles), such as may be seen today in recently active regions such as Iceland, Yellowstone and North Island, New Zealand (*Fig. 5.19*).

Fig. 5.19, below Hot springs near Rotorua, North Island, New Zealand

Chapter 6

Early Cainozoic Volcanoes:
the Big Ones

Fissure eruptions, with subordinate associated basaltic shield eruptions, led the way in the establishment of the Hebridean volcanic fields and were a fundamental aspect of the two to three million years of the early Cainozoic (Palaeogene) activity. Intense in the earliest stages, this type of activity prevailed to the latest stages as long as the extensional stresses on the lithosphere were exerted and sub-lithospheric mantle temperatures remained high. However, whereas the fissure and shield volcanoes would have been fairly minor excrescences on the landscape, much greater accumulations of lavas and fragmental deposits were piling up at favoured localities to form major volcanoes. Subsequent uplift and prolonged erosion has revealed their internal anatomy in the mountains of Arran, Mull, Ardnamurchan, Rum and Skye. The volcanic significance of these districts was clearly recognised well over one hundred years ago by Judd and Geikie. Other heroic names indelibly associated with the geological pioneering of these areas include Harker (Skye and Rum), Tyrrell (Arran), Bailey, Clough and Wright (Mull) and Richey and Thomas (Ardnamurchan). In the past sixty years many other resolute souls have endured the wind, rain, bogs, midges and other mortifications to the flesh to unravel the geological secrets of these areas.

Geological mapping in recent years of the sea floor west of the Hebrides and out to the margins of the continental shelf has revealed the sites of many other large central-type volcanoes, so that what we can actually see above sea-level is only a part of a much broader picture. Far out to the west the lonely sea-stack of Rockall is all that remains above the water of one such volcano. Closer to the Outer Hebrides the islands and sea-stacks of St Kilda are all that is visible of another. In Ireland, the Mountains of Mourne, Slieve Gullion and the Carlingford area mark the sites of roughly contemporaneous volcanic centres. Smaller volcanic centres may be represented by the early Cainozoic granitic intrusions forming the islands of Ailsa Craig and, far to the south, Lundy Island in the Bristol Channel. The whole upper and outer carapaces of these volcanoes have been stripped away by water and ice so that much of the internal anatomy is revealed for investigation. Consequently what we see in the major centres are masses of intrusive igneous rocks formed from magmas crystallised within the volcanic cores. Occasionally, as in Mull and central Arran, small portions of the original volcanic superstructure have been dropped down by caldera faults by hundreds of metres from their original elevation, thus saving them from erosional destruction.

On the geological maps the intrusive cores (plus the occasionally preserved extrusive rocks) tend to have roughly circular or ovoid forms with diameters of 5–12 km. Although the igneous rocks constituting these cores provide the factual data, envisaging the morphology, compositions and eruptive histories of the volcanic edifices that would have overlain these in the early Cainozoic is a matter for the imagination. The imagination, however, is not unconstrained but is kept within limits by the great number of studies of volcanoes worldwide showing all stages of evolution from their constructive and active phases, essentially untouched by erosion, to those that have been reduced to stumps.

In the past few decades forensic scientists have developed the skills necessary to reconstruct the muscles, skin and facial features to generate facsimiles of the dead, on the sole basis of a skull. Anthropologists and palaeontologists likewise work to produce the best approximations they can of early man and animals from skeletons, often fragmentary. In the same spirit, conceptual reconstructions of the long-gone Scottish volcanoes are attempted in these pages. In this and the following chapters I have tried to base these reconstructions on analogous cases. Thus we can guess at the original dimensions of the volcanoes by analogy with modern volcanoes of similar nature. Large continental volcanoes of roughly similar composition and structural setting reach heights of 3–4 km and extend over diameters of 50–75 km. If we assume similar dimensions for the early Cainozoic Hebridean volcanoes, much of Skye, Rum, Ardnamurchan and Mull would have been buried beneath these structures and there would have been a considerable degree of overlap between adjacent volcanoes.

Before going on to consider any of the major Hebridean volcanoes in detail, I invite the reader to imagine a vertical slice through a generalised, hypothetical example of one of these big structures in the prime of its eruptive life. The anatomy is complex: on the outer shell of our imaginary volcano we may expect to see lavas and intervening pyroclastic strata dipping outwards and away from the principal crater. Depending on the viscosity of the lavas and/or the angle of rest of any ashes from the latest eruptive phase, the flanks are likely have had slopes of between 20° and 40°

from the horizontal. In the vicinity of the crater(s), however, dips may be reversed, with lavas and debris flowing or tumbling back towards a conduit that had been cleared by the previous release of explosive gas. In the upper part of the structure intrusive sheets of igneous rock would be present, usually no more than a few metres thick. Although these may have very varied orientation they may be divided into those that: (a) are essentially vertical dykes, cutting across the lava and ash strata; (b) lie roughly parallel to the strata (sills); and (c) cut across the strata obliquely. These latter sometimes have a tendency to dip inwards towards a focus (or foci) deep beneath the summit of the volcano and, where they extend circumferentially around a substantial part of the volcano, they may have a generally conical form. Large numbers of such centrally inclined sheets can occur, forming a 'cone-sheet swarm' directed downwards and inwards to a focus at a depth of a few kilometres (*Fig. 6.1*).

As with the dykes, the majority of magma injections rising along propagating conical fractures would fail to breach the surface, although a minority would attain surface level and produce small, parasitic volcanoes on the flanks of the parent structure. Proceeding downwards through our hypothetical cross-section, the concentration of the assorted dykes, sills and cone-sheets would increase at the expense of the lavas and pyroclastic extrusive rocks produced at earlier stages of growth of the volcano and, at two or three kilometres depth, these 'minor intrusions' may become the dominant component. By repeated intrusion over a long period substantial volumes of the interior of the volcano may come to consist of almost 100% intrusive sheets. Descending further in our cross-section we pass beneath the base of the volcano pile or superstructure and into the underlying rock formations. While we can still expect to find dykes and cone-sheets in

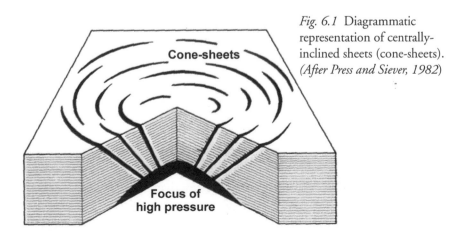

Fig. 6.1 Diagrammatic representation of centrally-inclined sheets (cone-sheets). (*After Press and Siever, 1982*)

ever-increasing amount, the number of large intrusions will also rise, representing locations where major volumes of magma were emplaced. Some will represent 'dead-ends', where magma ascended no further but cooled and crystallised in place. Others, however, will represent 'holding tanks' or magma chambers which, while intermittently receiving influxes of new magma from below, are concomitantly releasing batches of magma either to higher-level intrusions or to supply surface eruptions.

Much of the Hebridean region had, in the Mesozoic Era preceding the Cainozoic volcanism, been a region inundated by shallow seas on whose floors sands, muds and the shells of sea-creatures accumulated. The resultant sandstones, shales and limestones formed a near-surface barrier through which some of the early Cainozoic magmas were unable to penetrate. The downward pressure of the overlying rocks was insufficient to force these magma batches to the surface – much as low atmospheric pressure during a depression will cause the mercury to rise only a limited height in the barometer. Instead of rising higher, the magmas spread out laterally between the mechanically weak layers ('bedding planes') in the sedimentary rocks, forming intrusive sheets essentially parallel to the rocks above and below. Such sheets are referred to as sills. A beautiful example, seen in the Kilt Rock in northern Skye, is part of a major sill or cluster of sills underlying the Trotternish Peninsula (*Fig. 6.2*).

North of Skye the Shiant Islands represent part of this suite of sills, from which most of the overlying rocks have been stripped. The ruins of Duntolm Castle on Skye lie over part of the same sill. Another sill in the Mesozoic strata, easily visited on Skye, forms low sea-cliffs at Broadford. Elsewhere similar sills can be found, but intruded into still older sedimentary rocks. Thus, on Arran sills intruded into Permian and Triassic sandstones can be seen at Drumadoon, Dippin Head and Clauchlands Point. Still further south, on the north coast of Antrim, the great cliffs at Fairhead are due to a massive dolerite sill intruded into Carboniferous country-rocks.

As noted above, rising magma may become increasingly channelled into preferred conduits offering easier passages to the surface or near-surface intrusions. Locations where the dyke systems intersected pre-existing fault planes appear to afford such easy pathways. Magmas may be arrested at various levels within the crust, particularly at levels where the crustal rocks alter in composition and density. The resulting magma bodies, or chambers, may adopt many forms. Commonly they expand laterally to form lens-shaped chambers. Magmas are chemically reactive and may 'eat away' and 'digest' the confining rocks. As they crystallise they also give off heat, and since the basalt magmas are emplaced at temperatures over 1,000°C they

Fig 6.2 Sill, Trotternish peninsula, Skye. Sedimentary strata of Jurassic age intruded by the sill are seen in the foreground and in the right, background.

can melt the adjacent crustal rocks that commonly have melting points around 750° C, thus forming compositionally distinct secondary magmas, typically of granitic composition. Being lighter than the basaltic magmas that gave rise to them, they may stay as separate overlying magma bodies. However, evidence of mixing between basalt magmas and the crustally-derived magmas is frequently encountered.

The lifetime of the big Hebridean central-type volcanoes remains a matter of conjecture. However, we see in the eroded core complexes that hundreds, or even thousands, of separate intrusive events were involved. This is particularly so for the Mull, Ardnamurchan and Skye cases where, together with the larger slowly cooled intrusions (often referred to as 'plutonic intrusions') there are great numbers of the smaller intrusions, especially of the centrally-inclined cone-sheet type. Again we do not know the time intervals between one episode of magma admission and the next but these intervals are likely to have been of the order of tens to hundreds of years. The total time span for the growth of, for example, the volcanic structure over the Mull central complex may have been anything up to half a million years.

In the following section I attempt to give brief synopses of the central-type igneous complexes exposed on Arran, Mull, Ardnamurchan, Rum and Skye. The reader is reminded that there are many others, either wholly beneath the shallow seas of the continental shelf (like the Blackstones Bank complex, some 60 km WSW of Mull; *Fig. 5.7*) or only partially exposed, as in St Kilda, so that our knowledge is very limited.

Arran

There are large numbers of magmatic intrusions of early Cainozoic age on the Isle of Arran, all dating from a relatively short span of activity between 60 and 58.5 Myr. By far the largest and most prominent is the northern granite mass, out of which the Pleistocene glaciers have carved the scenic mountainous terrane of Goatfell, Cir Mhor and others. On the geological map this mass forms a sub-circular area about 12 km in diameter (*Fig. 6.3*). It consists of an earlier coarse-grained granite and a younger finer-grained granite. Rhyolitic (granitic!) magma, probably originating in the lower crust, took advantage of the plane of weakness provided by the Highland Boundary Fault. The early magma ascended and swelled out, balloon-like, in the upper crust, uplifting the overlying rocks. It thrust aside the wall rocks and severely distorted the Highland Boundary Fault which, by this stage in its history, was to all extents and purposes inert. Rapid erosion of the up-domed overlying rocks would have accompanied uplift. Although much of the space for the rising magma was generated forcefully as the pre-existing rocks were shoved aside it is possible, if not probable, that a sub-cylindrical mass of the pre-existing rocks subsided within the magma which flowed up around and over it. The mechanism by which the younger granite intruded, somewhat irregularly, into the older granite is unclear, but it did not deform the surrounding rocks in the manner of its predecessor. Although there is no doubt that these granitic magmas reached to within one or two kilometres of the contemporary surface, we do not know if they broke the surface to yield a rhyolitic volcano. If they did, it may be reckoned that an impermeable cap of solidified rocks formed sufficiently quickly to retain much of the volatile content of the early magma, permitting the crystals to grow to a size of several millimetres.

South of the northern granites, forming the less scenic high ground around Ard Bheinn (just south of 'the string road' across the island) is a second, sub-circular mass of intrusive and extrusive rocks (*Fig. 6.3*). This is the Central Ring Complex of Arran, so called (i) because it occupies a central place geographically, (ii) the component rocks describe a crudely

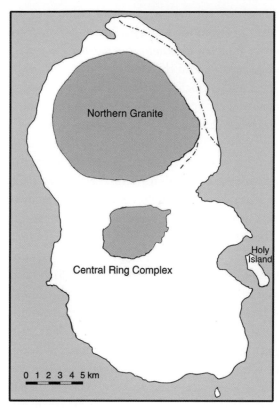

Fig 6.3 Map of Arran. The fault lines shown relate to the deformed Highland Boundary Fault. *(From Bell and Wilkinson, 2002, after Tomkeieff, 1961)*

concentric, annular pattern of discontinuous rings, and (iii) the relationships of the numerous types of rock composing it are certainly complex. The central complex is smaller (about 5 km diameter) than its northern neighbour and, because it cuts through the domed structure produced by the early northern granite, the central complex is demonstrably younger than the latter. Both were created within a time-span of around one million years. Although its emplacement also involved rhyolitic magmas that caused some up-doming of the adjacent country-rocks, this was much less dramatic than that brought about by its northern neighbour.

The change of activity from the northern granite mass to the Central Ring Complex involved a southward shift in the focus of magmatism of approximately 10 km. Such a migration of focus is very common in the evolution of volcanoes. Whether the northern granites crystallised beneath a volcanic edifice of their own making or whether the magmas never produced any surface eruptions remains an open question, but there can be little doubt that the Central Ring Complex represents the eroded core of a central-type volcano. This certainty arises from the observation that the early rocks of the complex include lavas and pyroclastic formations that had erupted at the surface. The lavas range in type from andesites through to rhyolites. Intrusive rocks (mainly granitic, but with gabbros and other types regarded as due to mixing or hybridisation between basaltic and rhyolitic magmas) tend to be younger than the extrusive lavas and pyroclastic rocks. The occurrence of early extrusions and later intrusions is explicable by postulating that the latter crystallised within an extrusive edifice that has since been eroded away.

A feature of particular interest in the Central Complex is that large masses, tens to hundreds of metres across, of older rocks are preserved within it. Many of these can be matched with the materials forming the country-rocks around the complex. These comprise coarse conglomerates from the lower Devonian 'Old Red Sandstone' formation and finer sandstones from the Permian–Triassic 'New Red Sandstone' formation. Since these form much of the outcrop in south and central Arran their presence as incorporated blocks within the volcanic complex is of only very mild interst. Much more importantly, there are, in addition, blocks of sedimentary rock whose fossil content shows them to be of Jurassic age. Other rock masses are of basaltic lava, some in contact with and overlying chalk limestone of Cretaceous age. The basalt blocks range up to 100 m thick to and 1 km long. Whereas Jurassic and Cretaceous (i.e. Mesozoic) rocks are not seen elsewhere on Arran they can be closely matched with the Jurassic and Cretaceous strata overlying the Permian–Triassic 'New Red Sandstone' some 80 km to the south-west in Antrim. Furthermore the basalt masses undoubtedly correlate with the early Cainozoic lavas that outcrop over most of Antrim.

Much of Antrim lies within the south-westerly extension of the Scottish Midland Valley, which was a lowland basin during Mesozoic and Cainozoic times. It may be inferred that Mesozoic and Cainozoic formations were originally much more widespread than they are today and covered Arran south of the Highland Boundary Fault, possibly extending still further north-east into the Midland Valley of the Scottish mainland. These inferences can only be drawn because of the chance preservation of the blocks in the Arran Central Ring Complex by down-faulting within the volcano. This dropped them to a level sufficiently low to escape the erosion which erased these rocks elsewhere on Arran and the Scottish mainland.

Detailed work on the central complex has allowed us to visualise the volcano. As noted above, the rising magmas created uplift and doming of the country-rocks, mimicking on a smaller scale what had accompanied the earlier ascent of magmas in the northern granite intrusions. Sub-circular fracturing occurred in the stressed roofing rocks above the magma, allowing violent escape of gases from the 'uncorked' chamber. Blobs of

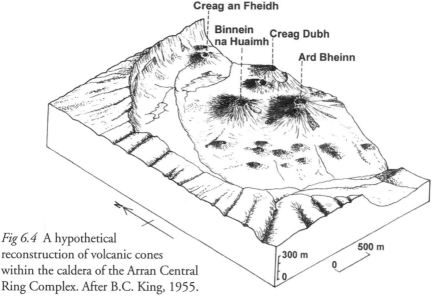

Fig 6.4 A hypothetical reconstruction of volcanic cones within the caldera of the Arran Central Ring Complex. After B.C. King, 1955.
Below, diagrammatic cross-section across Ard Bheinn (left) and Creag Dubh. *(After B.C. King, Quarterly Journal of the Geological Society, London, 1954)*

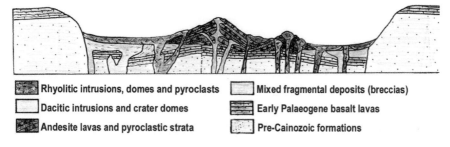

- Rhyolitic intrusions, domes and pyroclasts
- Dacitic intrusions and crater domes
- Andesite lavas and pyroclastic strata
- Mixed fragmental deposits (breccias)
- Early Palaeogene basalt lavas
- Pre-Cainozoic formations

magma and fragments of broken roof and side-wall rocks would have been expelled, building up accumulations of pyroclastic materials. Ash-flow and surge eruptions and the blanketing of the surrounding region by ash falling from eruption clouds probably accompanied this stage. Collapse of a roughly cylindrical mass of former roofing rocks within the ring fractures produced a pit or caldera whose diameter approximated that of the complex we see today. Within this a number of smaller volcanic cones grew and further localised up-doming took place within the caldera (*Fig. 6.4*). Such 'resurgent' doming after an initial major eruptive phase, giving rise to the formation of a caldera, is well known in younger volcanoes. The present hill tops within the complex may correspond closely to the individual volcanic foci within the caldera.

The whole volcano whose root zone is represented by the Central Ring Complex was probably at least 30 km across, covering all of what is now

Arran and much of its surroundings. If, as suggested above, the northern granites are also sub-volcanic, the Central Ring Complex volcano may have built up on the southern flanks of an extinct, fast-eroding bigger sister volcano. South of Arran the prominent island of Ailsa Craig is composed of microgranite of essentially the same age as those on Arran, at about 58.5 Myr. This microgranite, celebrated for the quality of its curling stones, forms a plug that may possibly also have fed a volcano.

Mull

The eroded carcass of the next great and relatively long-lived central-type volcano to be discussed is that on the island of Mull (*Fig. 6.5*). The igneous complex crystallised within the heart of this underlies the mountainous region of eastern Mull (*Fig. 6.6*), although not including Ben More Mull, which lies outside the complex and is wholly composed of the early lavas. As with the rocks of central Arran, it can be described as comprising a central ring complex, slightly elongate NW–SE and stretching over about 20 km. Most of the rocks are igneous intrusions although some are relicts

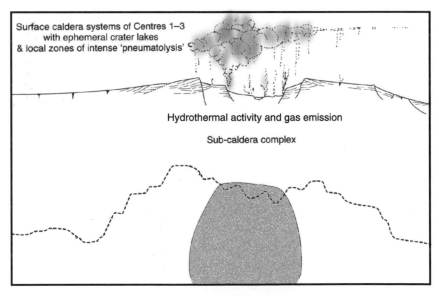

Fig. 6.5 A hypothetical section across the Mull volcano in one of its early, largely basaltic, growth stages. The figure illustrates a summit collapse pit (caldera) with associated hydrothermal activity (hot-springs and gas jets). Between the underlying core of coarsely crystalline igneous rocks and the outer shell of lavas, large numbers of minor intrusions, particularly cone-sheets, would be present. The dashed line signifies the present surface after erosion of most of the volcanic superstructure. (*After Bell and Wilkinson, 2002*)

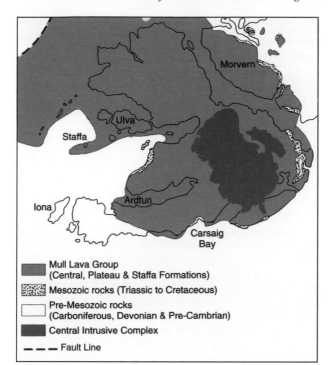

Fig. 6.6 Simplified geological map of Mull. *(After Bell and Wilkinson, 2002)*

of former surface lavas and pyroclastic rocks that escaped erosion as a result of having been dropped down within subsiding calderas. The intrusions include what are known as 'ring-dykes', i.e. dyke-like intrusions with steep (commonly outward dipping) contacts whose surface expression maps out as rings or partial rings (*Fig. 6.7*).

There are also thousands of cone-sheets, typically less than 3 m thick, dipping inwards and downwards to foci deep within the volcanic cores. Together with the ring-dykes, these inward-dipping sheets help to define the individual foci (*Fig. 6.8*). On Mull there are three such foci, representing the sub-volcanic cores of successive central-type volcanoes that were active in the later stages of its magmatic evolution.

The oldest of the central-type volcanoes overlay the Glen More focus (Centre 1) in south-eastern Mull (*Fig. 6.9*). The principal conduits beneath this presumably became blocked at some stage by congealed magma and new ones opened, slightly offset to the NW along the main zone of fissuring through which the early basalt lavas had erupted. A new volcano grew above this Beinn Chaisgidle centre (Centre 2) while its predecessor passed into extinction and decay. The Centre 2 volcano in turn became extinct and a third shift took place, with the focus of activity migrating still further to the NW to a point some 6 km from that of Centre 1 (*Fig. 6.9*).

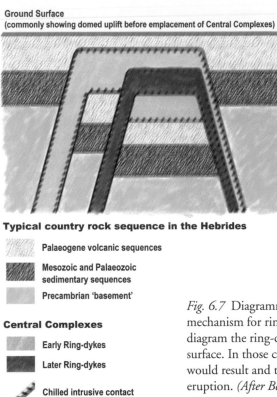

Ground Surface
(commonly showing domed uplift before emplacement of Central Complexes)

Typical country rock sequence in the Hebrides

Palaeogene volcanic sequences

Mesozoic and Palaeozoic
sedimentary sequences

Precambrian 'basement'

Central Complexes

Early Ring-dykes

Later Ring-dykes

Chilled intrusive contact

Fig. 6.7 Diagrammatic section showing mechanism for ring-dyke formation. In this diagram the ring-dykes do not reach the surface. In those cases where they do, a caldera would result and the ring-dyke would feed an eruption. *(After Bell and Wilkinson, 2002)*

Fig. 6.8 Hill-side near the head of Loch Scridain, Mull, composed almost entirely of cone-sheets relating to Centre 2.

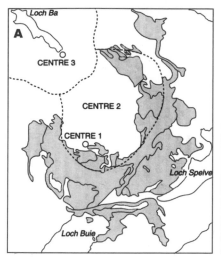

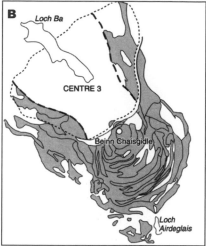

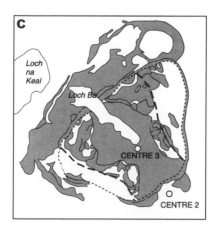

Fig. 6.9 Map of the three successive magmatic centres, Centre 1, 2 and 3 of the Mull Complex. A – Centre 1 (Glen More and the early caldera; B – Centre 2 (Beinn Chaisgidle); C – Centre 3 (Loch Ba and the late cladera. *(After Bell and Wilkinson, 2002)*

Investigators in the 19th century recognised that the Mull rocks represented an uplifted and sawn-down volcano, but it was not until 1912 that a team of young enthusiasts from the Geological Survey began to map the region in detail. The work, interrupted by the 1914–18 war and resumed thereafter, revealed an extraordinary complexity. The resulting map and memoir, published in 1924 by E. B. Bailey *et al.*, was in every sense a classic in the geological literature which opened up new vistas in the developing science of igneous petrology. Although recently republished by the Survey, the original 1924 map remains substantially unrevised. The story of the Mull volcano became famous throughout the geological community as a consequence of this seminal investigation.

On Mull, the evidence is more obvious than on Arran that, after a long history of basaltic eruptions supplied mainly through NW–SE-trending dykes, rising magmas became increasing channelled along conduits through the fractured rocks along a fault. As on Arran and in the cases of Rum and Skye to be considered below, we may conclude that hundreds (and possibly thousands) of cubic kilometres of basalt magma, liberated from the underlying mantle, accumulated in the deep crust. Fractional crystallisation of this magma led to continuous changes in its chemistry and a concomitant lowering of its density. The more buoyant magmas were then able to rise into the crust. Whereas on Arran an easy passageway was provided along a favoured site along the Highland Boundary Fault, magmas beneath Mull found comparable throughways at the intersection between the dyke swarm and the Great Glen Fault. Crustal melting in conjunction with fractional crystallisation produced derivative granitic magmas and these, as with their Arran counterparts, ascended and forced aside much of the pre-existing rocks in their path. The lateral forces exerted by the magmas caused the fault to become locally deflected by about 5 km. The radial compression also gave rise to a set of up-folds and down-folds (anticlines and synclines) in the disturbed country-rocks circumferential to the igneous complex. Subsequent erosion of this set of beautifully developed ring anticlines and synclines, affecting the very shallow crustal Mesozoic sedimentary rocks and early Cainozoic lavas around the Centre 1 intrusions, controlled the smoothly rounded (convex to the SE) coastline of the island.

The authors of the Mull memoir recognised that basaltic lavas of a type absent beyond the bounds of the complex had been saved from erosion by subsidence within an early Centre 1 caldera. Furthermore the form of these lavas (so-called pillow lavas, which will be discussed later in more detail) indicates eruption under water. This is likely to have been in a lake ponded during a period of repose within the caldera, like Crater Lake in the eponymous National Park in Oregon. The numerous intrusions (gabbroic, doleritic, microgranitic and with rocks of all intermediate compositions) in Centre 2 were mapped out as rings or partial rings in the 1924 survey. The probability is that many of these represent magmas intruded up annular fractures as repeated ring-faulting resulting in another caldera system.

It is, however, Centre 3 that provides the clearest evidence for caldera formation. A central block (or more likely an aggregate of blocks), some 18×11 km, sank into an underlying magma chamber. The subsidence has been estimated to have been as much as 900 m. The surrounding fault ranges from almost vertical to dipping outwards at angles of 70–80° so that as the central block went down space tended to open between it and the outside wall (cf. *Fig. 6.7*). Thus a potential space opened, rather as in a cone valve, allowing magma to rise and occupy it, although violent escape of gases may have helped to widen the gap. The magma arose to create a ring-dyke which can be traced for almost the entire periphery of the caldera. This ring-dyke, up to 100 m wide, was described in the 1924 memoir as 'the most perfect example of a ring-dyke known to science'; there are still few that surpass it (*Fig. 6.10*).

Salic rocks, mainly microgranitic (so-called granophyres), are very abundant at the present surface level. Some fairly large gabbroic bodies including the Bheinn Buie gabbro within the Centre 1 intrusions and the Corra Bheinn gabbro in Centre 2 (*Fig. 6.11*) are, however, exposed and may have been parts of large sill-like intrusions emplaced at shallow depths (2–3 km down). Although gabbroic rocks are relatively scarce at the current

Fig. 6.10 The escarpment in the foreground is formed by the steep contact (dipping towards the viewer) of the Loch Ba ring-dyke, Mull, Centre 3. Loch Ba itself is seen in the middle distance.

Fig. 6.11 Gabbros of the Corra Bheinn intrusion, Mull, Centre 2, form the foreground.

erosion level, geophysical investigations have shown that denser rocks (almost certainly gabbros and peridotites) occur at no great depth beneath the complex. It has been deduced that a sub-cylindrical body composed of these dense rocks underlies the whole intrusive complex, persisting down to around 15 km. The less dense microgranites, rhyolites and associates form a relatively thin topping above these unexposed gabbros and peridotites. This provides a reminder that it was the basaltic magmas, from which the gabbros and peridotites crystallised, that were the heat sources providing the energy that drove the entire volcano. In all probability a very similar situation obtained in the Arran centres and it is one that we shall meet again in other cases described below.

Early fissure eruption of basalt magmas tended to become more and more localised to specific focal nodes, primarily dictated by the presence of the Great Glen Fault. At these foci the magmas accumulated to form sizeable chambers, probably mainly in the shape of flattened sills or horizontally discoidal bodies that could be a few kilometres across and some hundreds of metres deep. Heat given off from these produced subordinate bodies of salic magmas which, because of their lower density, tended to overlie the basaltic magmas. Eruption of these salic (mainly rhyolitic) magmas would have produced ash-flows, surges and air-fall ashes as well as viscous lavas

Fig. 6.12 The 'mixed magma' rock forming the Loch Ba ring-dyke, Mull, Centre 3. Coin 25 mm diameter.

that would have accumulated as rather steep-sided cones above the successive centres. Each episode culminated in caldera collapse as their shallow magma chambers evacuated in violent eruptions that left the tops of the cones unsupported. Of these cones almost nothing remains since most of their dominantly fragmental materials were ephemeral and rapidly eroded. Some, however, remain as in Scallasdal (north-eastern Mull), where some of the bedded ashes and ignimbritic pyroclasts survive, albeit poorly exposed. What we do see are the sub-surface and generally more crystalline products, namely the medium-grained microgranites (granophyres) and also the pale coloured rocks, usually as small sheet-like intrusions that the survey workers referred to as 'felsites'. Commonly these latter are fine-grained devitrification products from what were originally rhyolitic glasses. Some of the best of the granophyre bodies are seen in Centre 3, in the vicinity of Loch Ba, where they occur as partial ring-dykes.

The closing scene of the Mull volcano was enacted at Centre 3 with the formation of the superb ring-dyke referred to above. This, the Loch Ba ring-dyke, is not quite rich enough in silica to qualify as a rhyolite and the term 'dacite' is preferred. It is a flinty, highly fractured rock whose exposures show it to be very heterogeneous. Within its pale grey matrix 'swim' elongate, vermiform lenticles of darker rock whose composition verges towards basaltic (*Fig. 6.12*). The form of these lenticles indicates that they were included as liquid blebs rather than as pieces of solid rock. The ring-dyke thus represents a 'mixed magma', a cocktail of imperfectly blended dacitic and subordinate basaltic magmas. Clearly the two contrasted liquids were simultaneously available during the injection of the ring-dyke. The very fine grain size implies either rapid cooling or loss of dissolved volatiles. The latter is the more likely and we may envisage a spectacular 'ring of fire', as magma fountained up all around the widening fracture that accompanied the sinking of the central mass. Dense clouds of gas charged with magma globules (pumice fragments) would have flowed into the growing caldera depression as 'ponded' ash-flows, as well as cascading radially outwards down the external flanks of the latest Mull volcano cone.

We know from the mixed nature of the Loch Ba ring-dyke as well as from many supporting lines of evidence that basaltic magmas were present at the same time as the derivative salic magmas. Investigations have shown that, in the closing eruption, the dacitic magma must have overlain the basaltic magma immediately prior to the collapse of the caldera, and that it was the extremely energetic uprise of both magma types during the collapse that led to their intermixing. At this stage the extensional forces acting across the Hebridean region were almost, but not quite, totally expended. This is demonstrated by the fact that a small number of NW–SE-trending basaltic dykes (implying continued lithospheric extension) were still able to post-date the Loch Ba ring-dyke and mark the finale of magmatic activity. The early Mull basalt eruptions commenced around 61 Myr ago and magmatism closed with the Centre 3 activity some four and a half million years later.

Ardnamurchan

Some 20–25 km NNW of the three Mull volcanic centres, another complex set of, mainly intrusive, igneous rocks forms the western end of the Ardnamurchan Peninsula. The rocks are generally similar to those of the Mull centres and, like the latter, they are regarded as the eroded cores of three successive central-type volcanoes. As on Mull they post-date the eruption of the early, near flat-lying basaltic lavas that outcrop over much of Mull and the Morvern Peninsula north of Mull.

We have seen that the siting of the first Arran centre was dictated by the Highland Boundary Fault and that the early Mull centre was similarly control-led by the Great Glen Fault. While there is no very obvious fault which may have influenced the siting of the Ardnamurchan centres it is possible that the magmas took advantage of fracturing associated with the Moine thrust belt. The term 'thrust' relates to low-angled faults (unlike the essentially vertical Highland Boundary and Great Glen Faults), where rocks have been pushed up and over underlying rocks. The Moine thrust belt separates rocks to the east, deformed in the Caledonian Orogeny, from the unaffected Foreland rocks to their west (*Fig. 2.1*). The thrust fault is traceable SSW from near Durness on the north-west coast of Scotland for nearly 200 km to the Sleat Peninsula of Skye. South of Skye its course is not precisely known but, if extrapolated a further 40 km SSW, the outcrop of the thrust faults could well have traversed western Ardnamurchan. If they did, all evidence has been obliterated by the much younger Cainozoic magmatism.

No high-precision radiometric ages are yet available for the Ardnamurchan rocks but, from their general geological relationships, there is every reason to suppose that the Ardnamurchan central volcanoes were broadly contemporaneous with those on Mull. On Mull, as we have seen, there was a consistent migration of the magmatic foci towards the NW but on Ardnamurchan there was no such consistency: the curvature of the early (cone-sheet) intrusions suggests an initial focus about half-way between the north and south coasts, NNE of Kilchoan. Following the demise of this, there was a relocation of some 5 km to the west, where Centre 2 became established. Finally, when the second centre exceeded its shelf-life, the volcanic focus switched back to a position roughly midway between those of Centre 1 and Centre 2 (*Fig 6.13*). *Fig. 6.14*, a geological relief map of Ardnamurchan, shows the roughly concentric ring pattern expressed both by rock-type and topography.

The products of Centre 1 in the Ben Hiant area are those of a sub-aerial volcano and differ markedly from those of Centres 2 and 3 in being surface or near-surface products, whereas those of the latter include coarser-grained intrusive rocks crystallised at considerable depth. The inference from this is that the younger volcanic piles were very much larger and higher and are likely to have totally overlapped that of Centre 1. Following later Cainozoic uplift, erosion peeled away the covers to reveal intrusions that had crystallised deep within the interiors of the Centre 2 and 3 structures but it only minimally affected the smaller early volcano.

Centre 1 mainly comprises a sequence of fragmental rocks up to 200 m thick, and a large volume of younger doleritic rocks. The sequence is poorly stratified and is generally unsorted with regard to the nature of the rock types and the size of the constituent blocks. Many of the latter are identifiable as having come from the early Cainozoic basaltic lavas, Mesozoic sedimentary strata and Precambrian metamorphic rocks from the 'basement' beneath the Mesozoic rocks. Mixed among these are fragments of extrusive rock-types that are otherwise unknown *in situ*. Prominently these involve pieces of rhyolitic and dacitic welded ash-flows

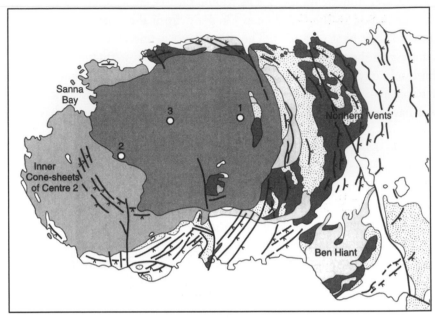

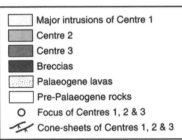

Major intrusions of Centre 1
Centre 2
Centre 3
Breccias
Palaeogene lavas
Pre-Palaeogene rocks
O Focus of Centres 1, 2 & 3
Cone-sheets of Centres 1, 2 & 3

Fig. 6.13 Maps of the three successive magmatic centres composing the Ardnamurchan Complex. *(After Bell and Wilkinson, 2002)*

(ignimbrites). These indicate that salic magmas erupted in the initial stages of the Ardnamurchan volcanic complex. It has also been observed that there was some marked up-doming of the country-rocks at an early stage. From these sparse pieces of evidence it may be postulated that, as in the two Arran centres and in Mull Centre 1, ascent of salic magma, probably largely derived from melting of crustal rocks, caused the doming and also gave rise to pyroclastic eruptions. To pursue this speculation, we may envisage the possibility of a caldera collapse on the summit of this hypothetical rhyo-dacitic volcano in the manner that we will encounter in the account (below) of the evolving Rum volcano.

The breccias of Centre 1 present a puzzle. While some of them could be the jumbled debris ('agglomerates;') left by the explosive degassing of the salic magma, some contain huge blocks, several tens of metres across, that

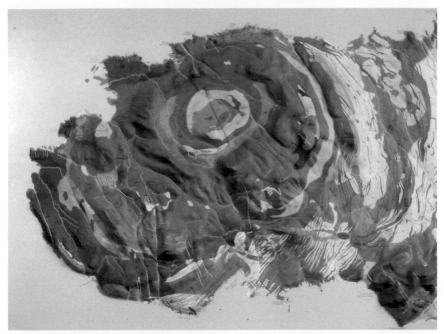

Fig 6.14 Geological relief map of the Ardnamurchan Complex showing that the remarkable annular geological features are well expressed in the topography. Colour key: pale cream (east side), Moine sedimentary rocks; yellow and khaki (south coast), Mesozoic sedimentary rocks; light pink (E and SE), basalt lavas; orange, Centre 1 fragmented rocks (breccias); dark grey around SW coast, hypersthene gabbro; the very prominent broad grey ring is the Centre 3 great eucrite enclosing other intrusive units of Centre 3.

Overleaf, Fig. 6.15 View of Ben Hiant, Ardnamurchan, from northern Mull.

can scarcely be imagined to have formed in such a manner. Recent investigation has suggested that much of these breccias resulted from the gravitational sliding and slumping of rocks down the volcanic flanks as domical uplift was taking place. If so, this uplift occurred, or at least continued, after rhyolitic-dacitic eruptions had taken place in order to account for the presence of the ignimbritic fragments in the breccias. Central uplift, caused by renewed ascent of magma after a caldera has developed, as described above for the central complex in Arran, may also have happened in Ardnamurchan Centre 1.

The early breccias are now distributed around the eastern and northern sides of the complex where they are beautifully exposed on the coasts. They are also well exposed on the southern side of Ben Hiant – although of dangerous access – in the steep cliffs of McLean's Nose. Several poorly

exposed bodies of glassy ('pitchstone'), slightly more mafic than dacite, occur in association with the breccias a short distance SW of the Ben Hiant summit. Although again subject to dispute, these may be lava flows possibly erupted within a crater environment.

The upper parts of Ben Hiant, the beautiful mountain that dominates the Ardnamurchan landscape (*Fig. 6.15*), are composed of sheets of dolerite that are younger than the breccias. Their total thickness amounts to several hundred metres and yet the magma (or magmas) did not cool to a coarse gabbroic product but to rocks whose textures imply relatively fast cooling rates. This invites the speculation that, late in the growth history of the Centre 1 volcano, basaltic magmas were successively ponded within a large crater or caldera and that what we see are rocks crystallised within 'lava lakes'. Lava lakes of long duration (tens to hundreds of years) are relatively uncommon phenomena. The most famous example in the last two centuries is that of Kilauea volcano on the main island of Hawaii. Others are known from Nyiragongo in Kivu Province, Central Africa, and Mt Erebus in Antarctica. Another, which may be more nearly analogous to the former Hebridean volcanoes, is the Erta'Ala lava lake in the Danakil depression of northern Ethiopia. The answer is uncertain, but it is possible that the Ben Hiant dolerites originated within a late-stage lava lake environment and, like the preceding breccias, have undergone only relatively superficial erosion.

Whatever may have been the growth history of the volcano built over Centre 1, we may be sure that a great deal of critical evidence was obliterated during the subsequent creation of Centres 2 and 3. Ardnamurchan Centre 2 is defined very largely by mafic rocks, namely by two generations of cone-sheets (*Fig. 6.16*), the remains of what must formerly have been a thick and extensive gabbro intrusion and by some partial gabbro ring-dykes. The thick gabbro intrusion represents a slowly cooled basalt magma chamber: it is called the hypersthene gabbro on account of the presence within it of hypersthene, a species of pyroxene as well as the more typical pyroxene, augite. This gabbro has a crescentic form on the map, curving around the north, west and south of the peninsula from Sanna Point, to just north of Kilchoan. Near the lighthouse it provides the westernmost extremity of the British mainland. The external margin of this crescentic outcrop is the outward dipping contact of the gabbro against older rocks (Mesozoic sedimentary strata, basalt lavas and early Centre 2 dolerite cone-sheets) that were thoroughly baked ('thermally metamorphosed') by it.

When it was emplaced, the hypersthene gabbro probably had a circular plan and a dome-like form but, like the Ben Buie and Corra Bheinn gabbros of Mull, it probably owes its present shape to having been intersected and

Fig. 6.16 A cone-sheet intruded into Precambrian meta-sedimentary strata, near Kilchoan, Ardnamurchan. This is one of the earlier Centre 2 cone-sheets that pre-dated the hypersthene gabbro.

cored-out by the eccentric emplacement of younger intrusions. A second dense swarm of doleritic centrally-inclined-sheets (the 'inner cone-sheets') post-dates most of these. Although the individual intrusive sheets are rarely more than 6 m thick, the composite thickness of the pre- and post-hyper-sthene gabbro cone-sheets amounts to around 1300 m. Because of their inward dipping conical geometry, this implies that the rocks on their inner and upper side ('inside the cone') experienced a relative uplift of a corresponding amount. Actual hydraulic uplift of over a kilometre is mechanically improb-able and much or all may have been accomplished by successive subsidences of the rocks beneath. Although Centre 2 is dominantly composed of mafic igneous rocks (dolerites and gabbros), some salic rocks (microgranitic) were also involved and it is apparent that at times here, as in the Mull centres, both magmas types were simultaneously available for intrusion.

The third and final focus of activity on Ardnamurchan was around a point about one kilometre east of the settlement of Achnaha. There are a number of ring-shaped intrusions within one broad and annular mass of gabbro known as the 'great eucrite'. This is a tough, homogeneous and coherent rock (*Fig. 6.17*) that withstood glacial erosion sufficiently to form

Fig. 6.17 A surface of the coarse gabbro in Ardnamurchan Centre 3, referred to as 'the great eucrite'.

an upstanding doughnut-like ring of high ground, providing an outer frame to Centre 3. It is the great eucrite in particular that confers such a striking circularity to the topography of Ardnamurchan (*Fig. 6.14*). That the term 'ring-complex' is particularly apt is clear to any observer standing on the crest of the great eucrite ridge, and is strikingly brought out in aerial views.

The magnificent sub-circular basin enclosed by the great eucrite is sometimes described as an old volcanic crater. This interpretation is, however, mistaken. The original volcano (together with any craters or calderas it may have possessed) deduced to have overlain Centre 3 was probably already extensively reduced by erosion with two or three million years of its extinction. What we see now, as in the other large Hebridean volcanic centres, is the relatively deep root zone after removal of two to three kilometres of its superstructure, largely effected by the Pleistocene glaciers in the last million years. It is fortuitous that the inner rocks were comparatively easily eroded, whilst the resistant outer rim of the great eucrite was left as encircling high-ground (*Fig. 6.14*).

The term 'ring dyke' has been employed in the past to describe the Centre 3 intrusions and some of them may indeed be of this kind. Although a recent work suggests that Centre 3 is rather composed of a suite of intrusions whose contacts dip inwards towards a common focus, constituting a nested suite of funnel-shaped intrusions, this is anything but certain. Because of inadequate exposures it is very difficult to verify whether the contacts dip in, are essentially vertical or are even outward dipping, leaving abundant scope for argument and further investigation.

We may, however, surmise that the intrusions that we see in Centres 2 and 3 were formed deep in the superstructure of two large and successive volcanoes that overlay the ruins of the Centre 1 edifice. Whether the argument goes in favour of ring-dykes with vertical or outward dipping margins or conically inward-dipping sheets, it is irrefutable that the volcanoes must have been dominantly basaltic and probably crowned by caldera pits. These volcanoes in their prime would have risen high above what was left of the Centre 1 structure (*Fig 6.18*).

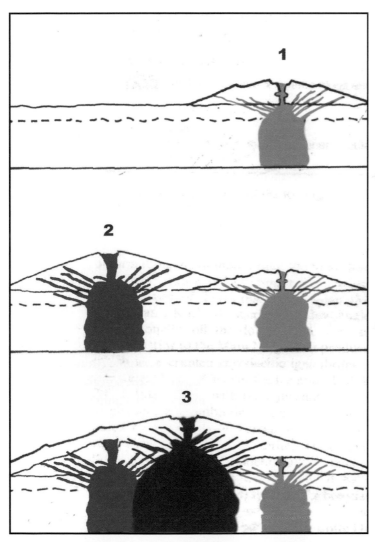

Fig. 6.18 Diagrammatic cross-section showing hypothetical evolution of successive volcanoes in Ardnamurchan. Dashed line shows schematically the present surface level after extensive erosion of the volcanoes.

Rum

From Ardnamurchan Point one can look southwards to the hills marking the Mull complex and north towards the rugged mountains of Rum and, beyond, to the Cuillins of Skye. The Ardnamurchan complex lies very roughly midway between the Mull and Rum complexes, approximately 30 km from each. In early Cainozoic times Rum lay on a ridge of relatively high ground that stretched roughly NNE–SSW from Skye southwards, including the rocks of Coll and Tiree. Magmatism on Rum may have been facilitated or focused by a fault zone that runs N–S, approximately bisecting the island (*Fig. 6.19*). This fault is called the Long Loch Fault after a prominent elongate loch in central Rum. Once again certainty is a luxury denied us but there are many indications that this fault, which may well have had a pre-Cainozoic history, was active during the formation of the igneous complex. The fault had right-lateral displacement(s), which means that to an observer, the rocks on the farther side of the fault have moved towards his right.

Despite the fact that there are many points of resemblance between Rum and the other big volcanic centres, there are also points of singularity that confer a very distinct character to Rum. Rum, like Mull, Morvern, Ardnamurchan, Muck and Eigg was inundated by the early basalt lavas for which fissure and fissure-related shield volcanoes are held to have been primarily responsible. On Rum, however, only vestiges of these flood basalts remain, most having been either obliterated by the subsequent magmatism or eroded. The Rum central volcano was inaugurated by uprise of salic magma and, once again, the principal source of this is inferred to have been melting of the underlying crust with basalt magma beneath it supplying the heat. Rum lies to the west of the Moine thrust zone and thus, unlike Ardnamurchan, is beyond the Moine meta-sedimentary rocks, within the so-called Caledonian foreland zone. In this foreland, the country-rocks comprise the ancient Lewisian metamorphic rocks (gneisses) that are overlain by a great thickness of red-brown sandstones. These are the tough, lithified products of sands carried down by ancient river systems from a now lost land to the west. The classic or 'type' area for these sandstones is the Torridon Forest area north-east of Skye and accordingly they are called the Torridonian sandstones. On Rum a sequence of these sandstones, over 2 km thick, forms a frame around the northern and eastern sides of the Palaeogene volcanic complex.

As on Mull, Ardnamurchan and over much of the Hebrides, some Mesozoic sediments were deposited. A thin layer of terrestrial sands and conglomerates of Triassic age overlies the Torridonian in northern Rum.

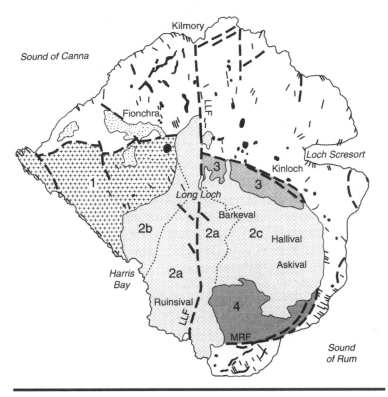

COUNTRY ROCKS

Palaeogene lavas (Canna Lava Formation)

CR Pre-Palaeogene country rocks

CENTRAL COMPLEX

1 Western Granite

2 Layered series (a-Central Intrusion; b-Western Layered Intrusion; c-Eastern Layered Intrusion)

3 Northern Marginal Zone

4 Southern Marginal Zone

MINOR INTRUSIONS

Peridotite, gabbro & dolerite plugs and mafic sheets

\\\ \ Dykes (selection only)

— — Faults
MRF – Main Ring Fault
LLF – Long Loch Fault

Fig. 6.19 Simplified geological map of Rum showing the intrusive complex. *(After Bell and Wilkinson, 2002)*

During episodes of heightened global sea-levels, shallow seas spread across the Hebridean basins in the succeeding Jurassic Period and shallow marine sediments were deposited. The Jurassic sedimentary rocks, however, like the early Cainozoic basalts that erupted over them, were almost (but not quite) destroyed by later magmatism or lost by erosion.

The uprise of the salic magmas that heralded the Rum central volcanic complex caused some outward dipping of the surrounding country-rocks but did not cause dramatic folding in the adjacent country-rocks as it did on Arran and Mull. A recent study of the early stages of the Rum volcano provides us with a detailed history (*Fig. 6.20*). The salic magma appears to have ascended in successive waves as more material was added incrementally to the growing magma chamber. The tumescence of the latter not only caused the surrounding rocks to dip outwards but also elevated masses of Lewisian 'basement' gneisses and strata low in the Torridonian succession above the apex of the chamber within a surrounding ring of faults (*Fig. 6.19*). This ring-fault was of some 15 km diameter around the developing central igneous complex and is traceable around the eastern side of Rum, passing about 2 km south of Kinloch Castle and westwards to the Sound of Canna. To the south-west of the island the trace of the fault lies beneath the sea. It is only because of successive uplifts, amounting in all to over 1 km, within the ring-faults that the Lewisian gneisses and lowest Torridonian strata can now be seen. Outside the bounding faults these older rocks lie buried beneath the younger Torridonian sandstones.

As tumescence continued, the apical region within the ring-faults experienced extensional stresses and began to subside into a basin or proto-caldera. As the basin walls steepened, rocks fell from them to form scree or talus deposits. These accumulations of angular pieces of rock were subsequently cemented together, typically by the action of percolating aqueous solutions, forming breccias. By the accumulation of large thicknesses of these talus materials, mainly composed of Torridonian sandstone, a pile of essentially unbedded breccias, up to 170 m thick, formed within the proto-caldera. These are easily observed as one enters Coire Dubh on the footpath from Kinloch Castle.

Doming, proto-caldera collapse and accumulation of sedimentary breccias provided the preamble to surface eruptions. At first these involved escapes of gas from the underlying magma chamber, limited in time and volume but forming thin pyroclastic horizons covering the growing breccia sequence. Having once opened, cracks permitting release of the pent-up volatiles widened and extended and the first major eruption involving magma ensued. This break-through may well have been triggered by heat

a) Salic magma chamber inflates above mafic magma (black), raising dome of older rocks.

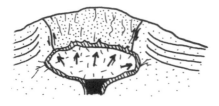

b) Collapse of dome: formation of proto-caldera and breccias.

c) New influx of mafic magma triggers eruption of salic magma.

d) Caldera collapse follows major eruption of salic magma.

Fig. 6.20 Evolution of the early salic central volcano on Rum. *(After V. Troll et al., 1997)*

introduced by the arrival of a new batch of basalt magma at the base of the salic magma chamber. We may guess that at this juncture the chamber roof underwent catastrophic failure: gas and pumice shot skywards while ground-hugging ash-flow sheets spread rapidly across the basin floor. As magma spilled from the underlying chamber further subsidence occurred with growth of the caldera. The magma, with a composition intermediate between that of dacite and rhyolite, erupted as a high-temperature (probably over 850°C) ash-flow. The glassy pumice blobs in the ash-flow settled out and compacted as the gases escaped, becoming flattened and welded together in the process and forming a tough coherent ignimbrite (*Fig. 3.8*). These rocks are now seen up to a thickness of some 80 m, for example in glaciated outcrops on either flank of Coire Dubh and in the southern walls of Dibidil in south-eastern Rum.

Although it has been estimated that some ten cubic kilometres of magma erupted at this time, this may well be an underestimate since the eruption products are only preserved in the eastern part of the large caldera, and even there the top of the ignimbrite unit has been lost to erosion.

Fig. 6.21 View of Hallival, Rum showing the terraced aspect that reflects differential erosion of the layered mafic and ultramafic rocks.

Underlying a large triangular area north of Harris and west of the Long Loch are moderately coarse granitic rocks referred as to as the Western Granite. This is superbly exposed in the high sea-cliffs to the north-west of Harris, and similar rocks outcrop in the east of Rum around Dibidil. The granites represent the intrusive equivalents of the ignimbrites crystallised from the sub-surface magma chamber and revealed as a result of deeper erosion. When considering the big salic volcano that marked the first stage in the build-up of the Rum centre, it is tempting to imagine that a not dissimilar volcano may once have overlain the northern granite of Arran, but of which all traces have since been eliminated. The case for a comparable early salic stage in the evolution of the Ardnamurchan volcanic complex has already been presented.

After the expulsion of some of the salic magmas, the basaltic magmas broke through to shallow levels as NW–SE-trending dykes and as cone-sheets which, extrapolate downwards to a focal point some 2.5 km below the present surface. These small-scale basaltic events could be considered as a prelude to the second stage of the Rum volcano. This saw basaltic magmatism on a much larger scale that generated a massive complex of mafic and ultramafic rocks, at the same time destroying a large sector of the preceding salic volcano. Although some of the products were gabbros, it is ultramafic rocks (peridotites), composed very largely of olivine, that dominate the complex and give rise to much of Rum's rugged scenery.

It is possible that movements along the Long Loch Fault zone opened up unimpeded conduits penetrating not only the deep crust but even the upper lithospheric mantle, through which copious volumes of basaltic magmas ascended. Here I remind the reader that, as outlined in Chapter 3, the first major mineral to crystallise from basalt magma is olivine and that progressive crystallisation of olivine reduces the amount of magnesium left in the liquid. Thus, as the magmas cool and olivine crystals separate from them, the magnesium content falls. On Rum it appears that the magmas post-dating the early salic volcano were distinctly magnesian and unusually hot: they may have been emplaced, and perhaps erupted, at temperatures approaching 1,200° C. Such high-temperature magmas reaching shallow levels without already having undergone extensive crystallisation *en route* are referred to as primitive. Their magnesian character conferred a great capacity for yielding olivine-rich (peridotitic) rocks.

Much of the scenic appeal of Rum is due to the group of mountains in the east of the island, namely Barkeval, Hallival, Askival and Trallval. They display remarkable stepped or terraced profiles attributable to the fact that they are composed of repeated layers of rock, typically some tens of metres thick, with variable resistance to erosion *(Fig. 6.21)*. While the bulk of each layer, forming the more easily weathered lower portions, is mainly composed of peridotite, the tops tend to be more resistant and craggier as a consequence of higher contents of plagioclase feldspar. While sixteen layers can be recognised in this mountain group, more must lie unexposed below and others doubtless originally lay above but have been lost by erosion.

Peridotite remains the principal type of rock in the mafic–ultramafic complex of the Central Intrusions *(Fig. 6.19)* around the Long Loch. Here too, although the separate layers are not so obvious to the untrained eye, the rocks did not crystallise from a single body of magma but from a sequence of magma pulses. The evidence grows that the magmas arose through a narrow channel or set of channels oriented N–S in the vicinity of the Long Loch Fault. The observation that there are no fine-grained 'chilled margins' between the successive intrusions indicates that each occurred while its predecessors were still hot. The structures in the peridotitic rocks composing the Eastern and Western Layered Intrusions are comparatively simple. They occur as thick sheets dipping gently, at angles typically no greater than 20° to the horizontal, westwards and eastwards respectively, towards the median Long Loch zone *(Fig. 6.22)*. However, in the Central Intrusion stretching at least 12 km from its northern extremity to where it disappears seawards around Papadil in the south, the internal structures are far more complex and have yet to be fully mapped and interpreted. This area, comprising the Central Intrusion (up to 4 km broad), appears to represent a section

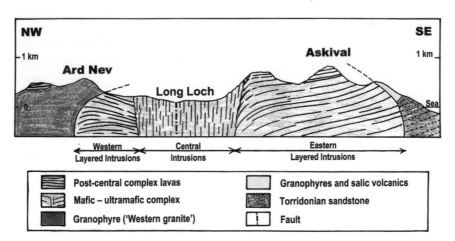

Fig. 6.22 Schematic cross-section of the Rum mafic-ultramafic complex showing disposition of the layering in the Western and Eastern Layered Intrusions, and the intervening Central Intrusions. *(After Emeleus, 1987)*

across a deep-rooted and extremely dynamic feeder zone which experienced both upsurges and down-draining of the basaltic magmas.

As noted above, at least sixteen such pulses of magma were inferred from the visible geology and the total number could have been at least twice this figure. It is suggested that the pulses ascended rapidly (thus retaining much of their heat and dissolved olivine component) up essentially vertical, dyke-like conduits until they reached a shallow crustal level where they spread out laterally as sub-horizontal magma bodies. These sideways leakages, which are thought to have taken advantage of the plane of weakness provided by the break ('unconformity') between the Lewisian gneisses and overlying Torridonian sandstones, appear to have been largely constrained by the ring-faults developed in the early phases of volcano growth. In these sill-like chambers, probably lying at little more than one and a half kilometres below the contemporary surface, the magmas cooled and precipitated much of their olivine component (producing peridotite) on the chamber floors. This was followed, in some cases, by crystallisation of plagioclase in addition to olivine, ultimately giving rise to the more resistant cliffs prominent on Hallival and Askival. A large proportion of the magma is deduced to have remained still uncrystallised at this stage. With its density reduced as a consequence of the copious separation of olivine to form the peridotite layers, the remaining magma was now capable of ascent still higher into the overlying volcano. One can consider these shallow magma chambers to have served as temporary holding stations or cisterns in which the magmas underwent cooling and partial crystallisation before final ascent and eruption. The chambers inflated with the arrival of new magma pulses and deflated with the expulsion of the previous batch of uncrystallised magma. The evacuation of the latter may have been triggered or driven by the arrival of new pulses of more primitive (hotter and denser) magma from the deeper mantle.

Much eruptive behaviour is pulsatory. The generation of the numerous layers of rock composing the ultramafic complex is but one manifestation of such behaviour. We do not know how much time elapsed between one pulse and the next, but the delays were long enough to allow crystallisation of tens of metres thickness of coarse (mainly peridotitic) rock in each layer. Although these may well correspond to intervals measurable in many hundreds of years, the rocks into which the sheet-like magma pulses were intruded remained, as noted above, hot enough to prevent formation of any finer-grained cooled contact material. Possibly each sheet settled deeper in the structure as a new arrival was intruded on top of it so that a stack of sheets formed, getting serially older from the top down. This hypothesis, however, remains contentious and will no doubt see modification as research continues.

After each new magma pulse flowed up the feeder zone there would have been some that was not injected laterally into the sub-horizontal chambers but, loaded with growing olivine crystals, sank back down the conduit up which it had ascended. That this is not entirely fanciful is supported by observations in active volcanoes, for example, Kilauea in Hawaii, where magma fountaining can be followed by relapse periods when magma swirls back down the feeder pipe. Thus although we generally think of magmas ascending through the crust, they can also retreat and descend like bath-water down a plug-hole. Tangible evidence for the vigorous movement of magma, either up or down, is provided by the profusion of blocks of already consolidated peridotite in the Central Series, ranging from tens of metres to centimetres across, occurring in disarray within matrices of younger peridotite. There are indications of down-dragging of still hot and ductile peridotites adjacent to the Long Loch zone that may have accompanied episodes of magma withdrawal. The dips of the peridotitic sheets inwards towards the presumed axial feeder zone may also relate to subsidence and magma withdrawal events along the axial zone.

The rocks now exposed in central Rum can be inferred to have crystallised at no more than 2 km depth beneath the surface at that time. The chemistry and mineralogy of the peridotitic rocks demand that not all of the magmas crystallised at the levels exposed and that a considerable proportion must have been erupted to higher levels. Eruptions probably took place along an elongate shield volcano lying north–south above the postulated Long Loch feeder zone. In the aftermath of eruptions, near-surface faulting is likely to have produced an elongate trench or volcano-graben. This may have resembled the volcanic gash in southern Iceland, Eldgjá, the site of a major eruption somewhat over a thousand years ago around the time of early Norse settlement (*Fig. 1.5*). A cross-section of the Rum volcano at this time may have resembled that of a shield volcano, with low-angled flanks and an axial eruptive trench. Despite some cooling and partial crystallisation in the shallow chambers the residual magmas would still have retained elevated temperatures on reaching the surface and were probably expelled as very fluid *pahoehoe*-type lavas.

It was suggested above that the time intervals between the formation of the numerous layered units in the mafic-ultramafic complex could have been measured in thousands of years. If the close relationship between layer formation and eruptions postulated above is true, then the repose periods between eruptions would have been of similar duration and contemporary erosion would have ensured that the volcano never attained any great height above the surrounding landscape. As with all volcanic systems where eruptive cycles are separated by long periods of inactivity, the structure would have been experiencing degradation almost as fast as it was growing. Consequently it was probably never able to accumulate more than 1 km thickness of lavas, passing down into a plexus of narrow feeder dykes traversing a former 'lid' of Torridonian strata. The latter would have lain directly above the coarse-grained layers eroded from those now forming the mountain tops of Rum. All of this lost 'roof', comprising ultramafic–mafic layered rocks overlying dykes and lavas, should then have been not more than 2 km thick.

Further speculations concern the relationship between the magmatism on Rum and the Long Loch Fault. The fault had probably been active in Mesozoic times and there has been some minor movement along it after the volcanic complex was emplaced. There is plenty of geological evidence from which to infer vigorous movements along the fault zone during the build-up of the mafic–ultramafic sub-volcanic complex which would have had seismic expression. In Chapter 1 it was explained how reduction of pressure on hot mantle rocks can trigger melting and it may be that episodic movements on the Long Loch Fault zone caused localised depres-surisation of the hot plume-head mantle beneath. Each time this happened the resulting high-temperature 'primitive' magma pulsed up the fault-generated passageways, replenishing the shallow magma chamber before the remainder was expelled at the surface.

In the hills of Orval, Fionchra and Bloodstone in the west of the island several lavas are seen, resting upon a surface eroded into the western granites formed in the volcano's early history (*Fig. 5.14*). Investigations have shown that these lavas occupied a succession of river valleys. As explained in Chapter 4, when a river valley becomes occupied by a lava flow, the river that previously ran through it will merely start to erode a new one, commonly on or alongside the obstructing lava. Hence one frequently encounters sections showing alternations – or oscillations – between lava and water-lain sands and gravels. Such is the case in western Rum, where a complex game was played between river erosion and volcanic filling over thousands of years. Many of the pebbles in these fluviatile deposits are derived from earlier lavas, granophyres or from the Torridonian sandstones. Some, however, are of coarse-grained gabbroic or peridotitic rocks that have come from the complex to the east. They thus provide proof that not only had the early salic volcano been deeply eroded but that this was also true for its mafic-ultramafic successor, within which these coarse-grained rocks had crystallised. The lavas post-dating the Rum central complex appear to have flowed from sources to the north and west, and correlate with the lava succession exposed on the Isle of Canna as well as those covering most of Skye north-west of Broadford and Loch Slapin. In the Skye lavas there are further examples of inter-flow pebble beds (conglomerates) deposited by rivers. Some of these contain granophyre pebbles matching the grano-phyres exposed on Rum. Since the next big sub-volcanic complex to be considered, namely the Cuillin, cuts through the Skye lava plateau, the age relationship between the Cuillin complex and its counterparts on Rum is established. Clearly by the time the rocks of the Cuillin were being crystallised deep in the interior of yet another great Hebridean volcano, the Rum volcano was cold and dead. It had not only been substantially destroyed by erosion but may have been, at least partially, smothered by the younger lavas flooding in from the north-west.

Skye

The fifth major Hebridean volcanic system to be considered is that of Skye. Here, both the craggy black Cuillin and the more rounded Red Hills are predominantly composed of coarse-grained intrusive rocks that, like those on Rum, crystallised at depths of no more than 2 km beneath large volcanoes. As with the Arran, Mull and Ardnamurchan examples, multiple intrusions tend to define specific foci or centres. On Skye, these migrated erratically over time through a distance of at least 15 km. Whereas in the case of the Rum volcano there was early uprise of salic magmas with a subsequent chaser of unusually hot basaltic magmas, the sequence was reversed in Skye.

The early plateau lavas that provide the distinctive scenery across almost all of the north of the island (*Fig. 5.2*) were, as discussed in the previous chapter, mainly erupted from fissure volcanoes and associated shields fed by the great NW–SE-trending Skye dyke swarm. As time passed the zone of melting in the mantle involved ever shallower levels and progressive increases in intensity. Such an expanding and shallowing evolution in the melting regime left an indelible mark in the chemistry of the basalt products. Not only were there compositional changes in the melt extract, but the magmatic productivity, in terms of unit volumes per year, probably increased. In the black Cuillin (*Fig. 6.23*) we see some of the rocks that crystallised from these more extreme melts as activity attained a climax.

It appears that the basaltic volcanism worked up to a crescendo over the passage of some one and a half million years and then diminished, termi-

Fig. 6.23 View of the Cuillins, Skye, from the north-west. Outcrops of the lavas predating the Cuillin Complex are seen in the foreground.

nating eventually at about 55 Myr. Furthermore the volcanism which, in its initial stages, had been widespread, became more localised around the area of the Cuillin. We have seen in earlier sections how magmas tend to seek pre-existing zones of weakness, particularly faults, to facilitate their ascent. On Skye it was a roughly north–south trending fault, the Camasunary Fault, that appears to have been exploited by the rising melts. The Camasunary Fault is an important structure that can be traced southwards from Skye, skirting the western flanks of Rum, Coll, Tiree and Skerryvore, and defining the western margin of one of the most prominent of the Mesozoic and early Cainozoic rifted troughs (*Fig. 5.7*). It falls into the category of a 'normal fault', with no sideways motion along it but with a relative down-drop on its south-eastern side. Although the idea that the Long Loch Fault on Rum had pre-Cainozoic displacements is debatable, it seems certain that the Camasunary Fault, a few kilometres to the east of it, was active during Mesozoic times. We may surmise that ascent of basaltic magmas in the Skye dyke swarm became concentrated in the vicinity of this fault. Volcanism thus evolved from predominantly fissure-type to central-type, culminating in a succession of great basaltic shields in the Cuillin area, built up above the earlier fissure-fed lavas.

These large and continuously evolving central-type volcanoes would have had gently dipping flanks surmounted by a succession of calderas which formed and filled sequentially. Magma accumulated within large cisterns or chambers not merely beneath the volcano, but within the edifice itself. Influxes of new magma from depth may have inflated these holding chambers but would also have promoted their emptying. The emptying would have taken place predominantly through eruption of lavas in the summit area, filling and over-spilling from the calderas and, increasingly as the volcano attained the maximum height that it was hydraulically possible for the magma to attain, by leakages through splits in the lower flanks. The volcanoes that formerly capped the black Cuillin probably resembled the large basaltic cone shown in *Fig. 6.24*.

It was probably through major flank eruptions that the high-level chambers underwent abrupt emptying. Each emptying left the summit area unsupported, resulting in collapse and caldera formation. The intervals between each of these major events were long enough for much of the magma in the chambers to crystallise as gabbros. What we see today in the Cuillin are the gabbros that grew successively within these shallow-level chambers, now revealed after the original superstructure had been stripped off by erosion. While almost every vestige of the lavas constituting the latter is gone, the geochemistry of the valley-filling basalt lavas at

Fig. 6.24 The caldera-crowned upper part of a large basaltic shield, Piton de la Fournaise, Reunion Island, Indian Ocean. Most is covered by ropy (*pahoehoe*) lavas, as seen in the foreground. Darker, rough (*aa*) lavas are seen in the distance. The summit regions of the big basaltic volcanoes of Mull, Ardamurchan and Skye were probably scenically similar. *(W.J. Wadsworth)*

Preshal More and Preshal Beg is such as to match that of the rocks within the black Cuillin. As mentioned in Chapter 5, these lavas flowed in to a river valley, deeply incised into the North Skye lavas, not far from Talisker. There is thus a reasonable chance that they provide vestigial evidence of the former black Cuillin volcano, some 12 km from the closest exposures of the gabbroic core.

The oldest components of the main volcano are the gabbros and associated peridotites of the black Cuillin *(Fig 6.25)*. Once again the term ring-complex can be applied since the outcrops of the numerous intrusions define a sub-circular pattern. The focus tended to migrate successively towards the ENE, but can be said to lie roughly beneath Meall Dearg at the southern end of Glen Sligachan. The contacts of the various rock units, together with mineral layering within them, generally dip downwards and inwards towards a common focus so that they compose a sub-parallel suite resembling massive cone-sheets.

The earliest and outermost gabbroic intrusion maps out as a broad frame, 1 to 2 km broad, around the northern, western and southern sectors from Glen Sligachan via Glen Brittle to Loch Scavaig. Another and somewhat younger voluminous injection of basaltic magma crystallised to form the partial ring of gabbros that forms an arc that is most accessibly seen around

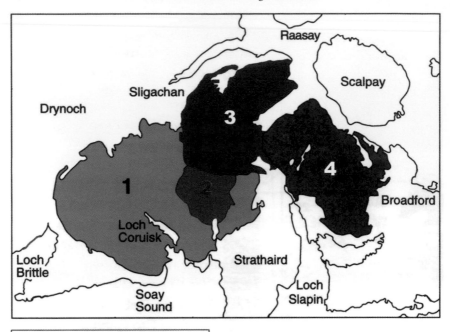

Fig. 6.25 A simplified map of the
principal units of the Cuillin Igneous
Complex, Skye. (*After Bell and Wilkinson,
2002*)

Fig. 6.26 View of Blaven from the south.

Loch Coruisk, and from whose easternmost outcrops Blaven (Bla Bheinn) (*Fig. 6.26*) and Garbh Bheinn have been carved by glaciation. The gabbroic rocks are not only coarse-grained (as they have to be by definition) but were highly resistant to the Pleistocene glaciers and, in consequence, they form some of the highest and most rugged topography in the west of Scotland, the celebrated Cuillin ridge.

Apart from the major (and therefore slower cooled and coarser) intrusions there are also great numbers of smaller sheet-like intrusions. Most of these are cone-sheets, rarely more than three metres thick, that were injected into the structure after solidification of the gabbros. Being quite small they lost heat relatively quickly to the cooled country-rock gabbros and crystallised as fine-grained basaltic or doleritic rocks. The cone-sheets tended to acquire close-set fractures or joints during and subsequent to cooling, a feature which led to their being more readily eroded and susceptible to plucking by frost action. It is largely the gaps and gullies where the cone-sheets have been worn away that create the serrated ruggedness of the black Cuillin, so endearing to the climbing community (*Fig. 6.27*)

As with the earlier examples described above from Mull and Ardnamurchan, intrusion of the cone-sheets involved either huge hydraulic forces, capable of elevating the rock mass above them, or subsidence of the rocks beneath them. The magma batches would have been injected rapidly, accompanying high-speed propagation of conical (or partially cone-shaped) fractures and each intrusion would have been accompanied by a seismic tremor as strain energy was dissipated. Three distinct episodes of cone-sheet intrusion have been recognised in the Cuillin. Volcanic earth-

Fig. 6.27 Sgùrr Alastair from Sgùrr nan Eag. Cuillin gabbros transected by cone-sheets (dipping upper left to lower right).

quakes related to this shallow-level fracturing and magma injection would have been frequent during the formation of the cone-sheet swarms.

Over the million or more years (approximately 59 to 58 Myr) during which basaltic magmas had been underlying and penetrating the crust a huge amount of heat was transmitted into the relatively fusible crustal rocks. This heating was, not surprisingly, concentrated in the area of principal activity, namely that of the future Cuillin. Rocks deep in the crust reached their melting temperatures and correspondingly large volumes of the essentially secondary granitic magmas were created. While the basaltic magmatism was coming to a close, these buoyant but viscous granitic magmas commenced their ascent to provide the final chapter in the volcanic evolution of Skye. We thus see on Skye the formation and demise of a great mafic volcano created from basaltic magmas, followed by an extensive history of contrasted salic volcanoes whose products were dominantly rhyolitic and dacitic pyroclasts and lavas and intrusive granophyres. Once again, later Cainozoic uplift of the crust was accompanied by deep erosion which wiped clean almost all of their surface products, now revealing the sub-volcanic ring complexes almost wholly composed of granophyres. Longonot, a caldera-topped trachytic volcano in the Kenyan Rift Valley *(Fig. 6.28)* has a form that may correspond reasonably closely to those of the former volcanoes that once overlay the Red Hills.

The earliest of the granophyres occurs below the eastern flank of Blaven in Coire Uaigneich. A larger sub-circular granophyre complex some 3 km

Fig. 6.28 Longonot volcano in the Kenyan rift. The general form of this volcano (largely composed of trachytic pyroclastic rocks) may be comparable to that of the salic volcanoes built up in the Eastern and Western Red Hills of Skye in the Palaeogene.

across, involving three separate intrusions south of Glen Sligachan around Srath na Creitheach, undoubtedly marks the intrusive core of a substantial salic volcano. An area of breccia (composed of sharp-edged blocks jumbled together and subsequently cemented together), pre-dates the granophyres. The broken blocks, up to several metres across, include basalts, dolerites and gabbros. It has been suggested that these blocks may have descended some 750–1,000 m from their original site of origin within a ring-fault. Some of the breccias were probably due to explosive release of gases from the rising granophyre magmas although others may represent debris flows of the kind common on the flanks of volcanoes, usually resulting from water-lubricated slurries of broken rock. Some may represent screes or collapse piles at the foot of a caldera wall. Bedded siltstones and ashy sand-stones also occur and the entire ensemble probably formed on the floor of an early caldera. Gravitational collapse was then followed by intrusion of the granophyres, themselves probably feeding surface eruptions whose products have been lost forever. The Srath na Creitheach complex thus has some features in common with the Arran central complex discussed earlier and it is specifically the early fragmental rocks in this complex that provide the clearest evidence that surface activity was involved.

The focus of activity next shifted north and east to the Western Red Hills area, between Loch Sligachan and Loch Ainort. Here intrusive suites of more or less concentric granophyre ring-dykes have taken a bite out of the NE sector of the earlier gabbroic complex and the Srath na Creitheach complex. It is the pinky-buff coloration of the granophyres, contrasting with the sombre greys of the gabbroic suite, that gave rise to the contrasting names for the Red Hills and the black Cuillin. The weathering characteristics of the granophyres are also in marked contrast to those of the gabbros, giving smoother slopes and offering little excitement to the climbing fraternity. The scarcity of cone-sheets also contributes to their less craggy nature.

Formation of the Western Red Hills involved some ten distinct introductions of magma, almost all salic, forming a suite of roughly concentric partial ring-dykes. Glamaig, the mountain rising just east of Sligachan, and Beinn Dearg Mhor are among the mountains formed from these. Several of these ring-dyke generating events may well have been accompanied by caldera formation, and a Western Red Hills volcano, composed mainly of rhyolitic/dacitic pyroclastic strata and subordinate lavas, was probably topped by a suite of nested caldera basins. A remarkable ring-dyke, up to 50 m broad, can be traced discontinuously through over 180° from the coast north of Loch Ainort west to Glamaig, south to Marsco and thence east towards Belig. The ring-dyke has as its principal topographic expression a conspicuous grassy gully carved out of the north face of Marsco. This is generally known as Harker's Gully in tribute to Alfred Harker, who first pointed out its remarkable geological features nearly a century ago. The ring-dyke is composite, involving both a salic component (rhyolitic) and an evolved iron-rich mafic component (ferro-diorite). The two magmas were available simultaneously at the time of the intrusion of the ring-dyke and there is clear evidence for their mixing as liquids. The ferro-dioritic component is notably rich in phosphate, in the form of the mineral apatite. East of Marsco, the soil produced by weathering of this rock and containing an enhanced phosphate content, makes the trace of the ring-dyke conspicuous as a result of the bright greeness of the vegetation.

There are various features in common between the Loch Ba ring-dyke on Mull and the Marsco suite ring-dyke on Skye. Both were mixed-magma intrusions, with the probable interpretation that, before their emplacement, the less dense salic magma overlay the denser mafic magma. Dramatic ascent to fill the opening ring-dykes led to more or less chaotic mixing of the two magmas. In each case, both sets of magmas congealed to very fine-grained rocks; the salic magma would initially have formed a rhyolitic glass which, as was so generally the case, subsequently devitrified to a microcrystalline rock.

As at Loch Ba, the Marsco suite ring-dyke was created by the subsidence of a plug-like body of rocks to its inner side (about 5.5 km diameter) and the magmas may be imagined as having reached the surface via a ring-fault to erupt as an arcuate 'fire curtain' around a developing caldera. Another analogous case appears to be present near Kilchrist, SW of Broadford and described below. A further close analogue to these mixed or hybrid ring-dykes has been described from the approximately contemporaneous Slieve Gullion complex in Northern Ireland.

Gabbroic rocks near the summit of Glamaig are also products from basaltic magma that co-existed with the rhyolitic magma from which the granophyre formed. This, in combination with the evidence from the Marsco suite, demonstrates that mafic magmas were still around in post-black Cuillin times even if their shallow-level manifestations were by this stage few and far between. Geophysicists investigating small variations in the gravity field across the region have concluded that the granophyric rocks of the Red Hills form only a relatively thin veneer, not more than 2 km thick, sitting upon much larger volumes of underlying dense rock that is almost certainly gabbroic.

The final stages of the great Skye volcanism saw the production of further salic volcanoes, with the focus of action now shifted to the south and east from the Western Red Hills complex just described. These youngest structures are those of the Eastern Red Hills, extending over a broad area from Loch Ainort south to Beinn an Dubhaich on the eastern side of Loch Slapin. Three successive intrusions of salic magma produced a suite of nested granophyre bodies, the youngest of which forms the mountain of Beinn na Caillich between Broadford and the head of Loch Slapin *(Fig. 6.29)*. The Beinn na Caillich granophyre has an almost perfectly circular plan, about 3 km in diameter and, apart from its north-eastern sector, is surrounded by fragmental volcanic rocks (volcanic breccias). These find their fullest development in a mass about 1 km broad on the southern flank of Beinn na Caillich, close to Kilchrist. The fragmental and partially bedded breccias are younger than the first two Eastern Red Hills granites but are clearly post-dated by the culminating Beinn na Caillich intrusion.

Detailed work on these rocks has shown them to be very important in assessing the palaeo-volcanology. The coarse deposits, which include blocks of both granophyric and gabbroic rocks, may be largely composed of debris resulting from the explosive release of gas from depressurised salic magmas approaching the surface. Some leached, lateritic horizons indicate sub-aerial weathering, although some ash beds appear to have been water-lain, probably within a 'crater lake'. There are also relicts of rhyolitic welded ash-

flows (ignimbrites) and lavas amongst the pre-Beinn na Caillich deposits. Like the earlier Srath na Creitheach rocks, these surface deposits around Beinn na Caillich owe their preservation to down-faulting within a ring-fault and provide the clearest evidence for the former presence of a caldera. Although the whole assemblage – including ash-flows, lavas, explosive fall-out debris and possible scree and debris-flow materials – has been partially preserved within a caldera, these materials would originally have had a very much wider geographic spread, and represent eruptive products from the

earlier granophyre magmas of the eastern centre. Had they not been saved from erosion by subsidence within a down-faulted block, attribution of the Eastern Red Hills to surface volcanic activity would have been wholly hypothetical. As in the central complex of Arran we see here materials that were once present at the surface intruded by a younger granophyric body. The Beinn na Caillich granophyre magma was itself admitted in a final act that involved down-dropping of a cylindrical mass of rocks within a ring-fault. Magma that arose along the sides of this and over its top would

Above, Fig. 6.29 Beinn na Caillich, Eastern Red Hills, Skye.

almost certainly have fed surface eruptions, and the granophyre now forming the summit of Beinn na Caillich may be presumed to have crystallised beneath a thick lid of its own extrusive products. Beinn na Caillich thus represents the eroded remains of the culmination of a long sequence of central volcanoes on Skye. Magmatism of the Eastern Red Hills has recently been dated to between 56 and 55.5 million years ago which still makes it earlier than the actual continental crustal separation that marked the onset of the North Atlantic Ocean.

Concluding observations

In Rum large-scale uprise of salic magmas gave an early rhyolitic/dacitic volcano, which was later swallowed up by the ascent of mafic magmas and the formation of a new volcano over the ruins of its predecessor. On Skye, the principal appearance of salic magmas was delayed until a late stage when the mafic magmas made only rare appearances. One can only guess at the reason for the contrast in timing of the mafic versus salic volcanoes at the two localities. On Rum basaltic magmas generated in the mantle were retained at depth in, or below, the crust for long preparatory stages, during which the heat evolved from the crystallisation of these deep reservoirs was transmitted to the overlying crust. The consequent crustal melting resulted in the growth of the early salic volcano. It may be that only after this volcano attained mature and inactive old age did the appropriate structural circumstances occur for the mantle melts to reach the shallow crust and thence the surface. On Skye, such deep crustal retention was probably less significant, and there was continuous heating of the crust by persistent flux of basalt magma until extensive crustal melting ensued. In the late volcanic stages on Skye when the Western and then Eastern Red Hills volcanoes came into being, the crystallising mafic magmas that constituted the heat engines remained almost entirely beneath the salic magmas. The latter may be thought of as the cream on the magmatic coffee!

Although the basaltic magmas, thought to have provided the driving agency below the Skye granophyres, generally failed to penetrate their salic magma cover they did, nevertheless make some high-level appearances. As we have seen, this happened at Glamaig and again in the hybrid rocks of the Marsco ring-dyke suite. Further tangible evidence that basalt magma was never far away comes from the rocks partially surrounding the fragmental rocks above Kilchrist. Here, in an intermittently developed ring-dyke, probably fed up along a ring fracture surrounding the collapse, is another suite of hybrid rocks formed by the incomplete mixing of the two contrasting magma types. In Centre 3 on Mull, where the culmination of volcanism almost exclusively involved salic magmas in what is thought to have been a classic caldera-topped central-type volcano, the extensional forces exerted by the convecting asthenospheric mantle deep below were not totally extinguished. As a result some further crustal splits along the well established NW–SE trend took place, admitting some very late-stage basaltic dykes, after the Loch Ba caldera had formed. On Skye also, the final death-throes of magmatism saw the intrusion of a few more small basaltic dykes.

In Chapter 5 reference was made to the extraordinary neo-glassy flow represented by the Sgùrr of Eigg. The light-house on the lonely reef of Oig Sgeir, some 30 km west of Eigg, stands upon almost identical glassy ('pitchstone') dacite and undoubtedly represents part of the same eruptive flow. As explained in Chapter 5, the Sgùrr flow is very thick and it is difficult to credit that a material of such high silica content as a dacite, with its correspondingly high viscosity, could have flowed a distance of 30 km or more. The observation that the Sgùrr (and Oig Sgeir) pitchstones contain a high percentage (over 30% by volume) of early-formed feldspar crystals makes the case for its having been a lava flow still harder to maintain. The presence of so much crystalline material would only have added to its sluggishness on the move. A possible answer to this conundrum is that it did not flow as a lava but as a high-temperature ignimbrite, i.e. a suspension of silicate pumice and crystals in a gas which constituted a lubricating medium. Such an ignimbritic ash-flow could readily have travelled the requisite distances. To take this hypothesis a stage further: if the ash-flow was sufficiently hot it could have continued to move in the manner of a lava flow after the gases separated and escaped, homogenising itself in the process to such a degree that the constituent pumice blebs lost their identities. Such a flow would be referred to as a 'rheomorphic ignimbrite', which superficially looks very like a normal lava flow. No source has been positively identified for the Sgùrr of Eigg–Oig Sgeir pitchstone. However, the rocks are dated to approximately 57 Myr and an intriguing line of speculation is that the flow might have originated from an eruption at the Beinn na Caillich centre, 40 km to the NNE on Skye. Again it would be ludicrous to postulate that a very crystal-rich dacite lava flow could have travelled so far, but the idea becomes more credible if it had spread far and wide as a catastrophic ash-flow. While this may be dismissed as a flight of fancy, the questions posed by the rocks remain unanswered and invite further investigations.

Although during their active periods the summit regions of the big Hebridean volcanoes would have been deserts floored by lavas and pyro-clasts (cf *Fig. 6.24*), the long repose periods would have seen their progressive colonisation by plants, of which ferns may have been among the earliest. It is improbable, with the heights they are likely to have attained and the warmer prevailing climates, that they were ever snow-capped. Over longer periods of quiescence the big volcanoes probably stood as fern- or shrub-decked uplands rising high above the lowland forests, their flanks scarred radially by river gorges.

Museum dioramas of the early Cainozoic tend to depict a landscape dominated by steep-sided volcanoes belching ash and gas, with lavas flowing down their sides. However, actual synchronicity of eruptions at the different centres is unprovable and, in any case, unlikely. Again it should be emphasised that long periods of repose, which one might think of as consisting of 'ground-hog days' when nothing much happened, inter-vened between one eruption and the next. Eruptions tend to be dramatic and geologically brief, lasting from days to perhaps a few decades, whereas the intervening periods would commonly have lasted tens to thousands of years. Bearing these comments in mind, the principal Hebridean volcanoes can be said to have been very roughly contemporaneous. Most were active from 61 to 58 Myr, with subordinate volcanism persisting for another three to four million years. The latter was contemporaneous with the much more productive volcanoes that came into eruption from around 55 Myr, to the north-west in Greenland and the Faeroes. Up to this time the embryonic Atlantic Ocean had yet to begin opening.

Chapter 7

Scotland within a Super-Continent: Upper Carboniferous and Permian Volcanoes

So far I have been considering the relatively recent, Cainozoic, volcanoes of Scotland and I now invite the reader to fast-track back with me another two hundred million years. This takes us across two great traumas for life on Earth – namely those used to subdivide the Cainozoic era from the preceding Mesozoic era and the Mesozoic from the Palaeozoic era. The Mesozoic extended back from 60 to 248 million years and the Palaeozoic from 248 to 542 million years. Just crossing the younger of these two time markers takes us into a very different world. Since at the start of the Cainozoic (i.e. the Palaeogene) there was no North Atlantic Ocean, the ocean and atmospheric circulation patterns would have been very different. If modern atlases were already useless to us in the Palaeogene, in these still earlier times they become utterly irrelevant, and we need to mentally adjust to the concept of a totally different distribution of land and sea. The Scotland we have now was non-existent, its landmass being subsumed within earlier continental configurations.

Through much of the Mesozoic, other than its earliest Period, the Triassic, the land-surface corresponding to what we now call Scotland and its environs would have been green, but without grasses and without the flowering plants and broad-leaved trees that were to become dominant in the Cainozoic. The vegetation would instead largely have comprised conifers, cycads and tree-ferns as well as ferns and more primitive land plants. Although ancestral mammals were around they were small and insignificant and the land fauna was dominated by reptiles, including dinosaurs. While there were some flying or gliding species, true birds had yet to make their appearance although feathers had already evolved on the 'dino-birds' of the Jurassic. Warm shallow seas to the west, east and south of what is now Scotland were home to fish and swimming reptiles and, among the invertebrate inhabitants, the ammonites were ubiquitous. These relatives of the squids and octopuses had coiled shells that were preserved in their millions as fossils in the compacting marine sediments. The nature of the catastrophe – or catastrophes – that took place around sixty million years ago that spelt extinction for so many families of plants and animals, largely wiping the ecological slate clean and laying the environment open for the explosive evolution of modern life forms, remains debatable. It is likely to have involved massive atmospheric pollution and a darkening of the skies as sunlight was blocked out, consequently spelling death to all but the most resistant plants and, of course, the animals that depended on them. Vast eruptions of basaltic volcanoes in what is now north-west India may have been responsible. Alternatively it could have been the arrival of a major extra-terrestrial body that impacted in the Yucatán region of Mexico or perhaps, more probably, a combination of these two events that produced the biospheric disaster.

During the three periods that make up the nearly 190 million years of the Mesozoic Era (namely the Cretaceous, Jurassic and Triassic) volcanoes left no Scottish record. Although pictorial reconstructions commonly depict volcanoes erupting in the distance while dinosaurs roam in the foreground, the dinosaurs of Mesozoic Scotland saw none, although there was some volcanism on the continental shelf west of Scotland and also not far to the east, in what is now the North Sea. Elsewhere in the world, as for instance in the American Cordillera, volcanism was occurring on a grand scale but the geology of the British Isles shows that there was no local activity.

Accordingly we shall skip back across the Mesozoic, crossing the doom-laden marker that separates the Mesozoic from the preceding Palaeozoic. The term 'doom-laden' relates to the fact that a huge percentage of all the plant and animal groups living at the end of the Permian (the last Period in the Palaeozoic) failed to survive into the Triassic. This quite horrific extinction was on a far greater scale than that at the end of the Mesozoic and the precise cause(s) are still more contentious. It appears not to have been instantaneous but to have been drawn out over several million years. Although the cause of this biological crisis remains enigmatic, one leading hypothesis links it to the vast outpourings of basaltic lavas in northern Siberia at this time. Dust and gas emissions from these may have been responsible for dramatically blocking solar radiation and for consequent lethal fluctuations in atmospheric temperatures. Whatever the answer, the event cleared the ecological deck for new forms of plants and animals including, most famously, the dinosaurs. Most particularly, our own ancestors (who appear to have been distinctly unattractive Permian reptiles) gamely managed to survive these widely fatal conditions. Whereas Chapters 4 and 5 considered a post-dinosaur Scotland, in this and subsequent chapters we have now reverted to pre-dinosaur times.

Throughout the Permian Scotland lay within a great continent to which the name Pangaea has been given. Pangaea came into existence as a super-

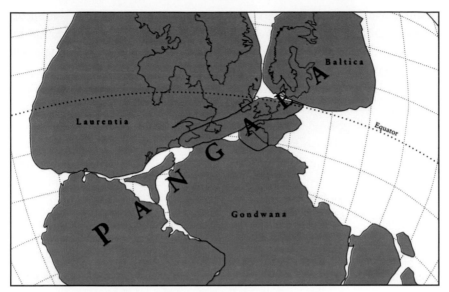

Fig. 7.1 Palaeogeographic reconstruction of the continents in the late Carboniferous, 320 m.y. ago. *(After D. Stephenson et al.,(eds.) 2003)*

continent as a result of plate movements that brought several continental masses into conjunction. The final continental docking that resulted in Pangaea occurred towards the end of the Carboniferous Period (*Fig. 7.1*).

The latest Carboniferous as well as the Permian and Triassic rocks of Scotland were formed deep in the interior of Pangaea and, being far from open seas and cut off from moisture-laden winds, the region experienced conditions of extreme aridity. The high ground, i.e. the ancestral Grampians and Northern Highlands, would have been rocky upland deserts while vast sand seas accumulated in some of the fringing lowlands. The region lay in tropical latitudes far to the south of where it does today and conditions would have been akin to those of the present-day Saharan or south Arabian deserts. Remains of the dune-bedded red sandstones deposited in these sand seas are beautifully exposed e.g. on the eastern coasts of Arran and in gorge-sections of the river Ayr near Mauchline. (Such sandstones make excellent building materials and were widely used in towns such as Carlisle, Glasgow and Edinburgh.) Because of the lack of fossils in these desert deposits there is no means of seeing when the end-Permian extinction took place. In the Scottish geological record it is nigh on impossible to identify any break between the Permian and Triassic strata. The desert deposits of the Permian and Triassic are accordingly collectively regarded as composing the New Red Sandstones (in contrast to the Old Red Sandstones that had accumulated in late Silurian and early Devonian times). Thus the end of the Palaeozoic and the start of the Mesozoic are hidden at some level within the New Red Sandstone formations of Scotland, which are mostly preserved in the Hebrides and south-western Scotland, with one small outcrop in the north-east near Elgin.

Volcanism in the Permian and later Carboniferous was almost entirely related to faulting and consequent pressure relief on the underlying mantle and can be categorised as wholly 'intra-plate' (see Chapter 1). Pangaea was no sooner formed than it began to experience extensional stresses leading ultimately to its disintegration in the Mesozoic and early Cainozoic. The magmatism tended to be low key and the volcanoes, although numerous, were small ones which did not give rise to any dramatic landscape features in Scotland. While we find their remains dominantly in the Midland Valley, particularly in the south-west (Ayrshire) and in the north-east, in Fife (*Fig. 7.2*), there was also scattered activity through the Hebrides and Western Highlands and also in Orkney. In the Orkney archipelago and extending onto the far north-east of the Scottish mainland there is a group of volcanic necks or vents, associated with some two hundred small mafic dykes. These are generally considered to be of late Permian age. If so they represent the youngest Palaeozoic magmatism in Scotland.

The volcanic vents, typically with diameters of only a few hundred metres, are commonly referred to as 'diatremes'. Diatremes frequently form when rising magma encounters water at shallow depths. The heated, trapped, water flashes into steam, throwing mud, rocks and lumps of molten magma high in the air. Often such diatremes show evidence of laterally-directed surge eruptions. The surface activity forms a wide crater bounded by a low rim of expelled debris, giving rise to the descriptive term 'tuff-ring', the name given to such broad circular craters surrounded by pyroclastic ramparts. The term tuff relates to the rock produced by consolidation of the pyroclastic fragments. Lakes that often develop within a tuff-ring are known as 'maars', a word originating from the circular lakes ponded within tuff-rings in the Eifel district of Germany. As the diatreme conduits clog up after each explosive pressure release more waters infiltrate the system to be re-heated, thus giving a repetitive sequence of explosions. Activity at such volcanoes is likely to be short-lived, lasting perhaps a few months, ending when the magmatic rocks become sufficiently cool and the volcano simply runs out of energy. *Fig. 7.3* shows a cross-section of an idealised diatreme.

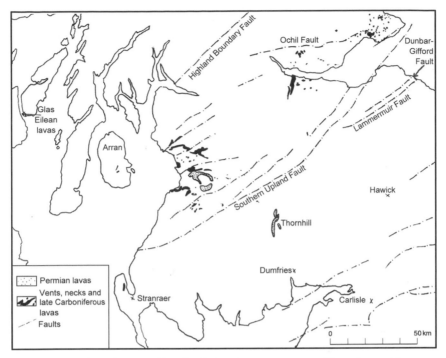

Fig 7.2 Map of south-central Scotland showing occurrences of Permian and late Carboniferous lavas. *(After W. A. Read et al., 2002)*

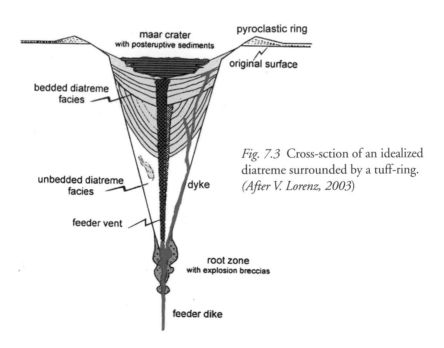

Fig. 7.3 Cross-sction of an idealized diatreme surrounded by a tuff-ring. *(After V. Lorenz, 2003)*

Ayrshire

Faulting in the earliest Permian and latest Carboniferous produced a series of rifted desert basins trending NNW–SSE across south-western Scotland, the North Channel and the extreme north-east of Northern Ireland. The rift faults localised ascent of basaltic magma, although on a very much reduced scale compared to what was going to happen 250 million years later in the Palaeogene. In Ayrshire, south of Kilmarnock and east of Ayr, numerous basaltic volcanoes erupted in this desert environment which had already been established by about 295 Myr, before the close of the Carboniferous. The lack of fossils in these strata makes it impossible to categorically ascribe an early Permian or a late Carboniferous age to the volcanism. The volcanoes are now represented by several dozen diatremes, mostly filled with fragmented rock. One example forms the prominent hill behind the town of Patna, while another can be seen at Waterside, approximately midway between Patna and Dalmellington. The greenness of many of these hills is attributable to the nutritious soils developed over the weathering basalt fragments, providing among other elements, higher concentrations of magnesium and phosphorus than are available from the Carboniferous sandstone country-rocks.

The pipes have experienced relatively little erosion and probably not more than around half a kilometre thickness of rock has been removed from their tops. So although the volcanic cones that originally overlay them are gone, a fairly substantial sequence of their lavas remains. The lavas, which form a succession some 240 m thick, are referred to collectively as the Mauchline lavas. These are directly overlain by, and interbedded with, the red desert sandstones; since the lavas and sandstones were both subsequently down-warped into a basin-shaped down-fold (the 'Mauchline Basin') and later eroded, the lavas are now seen as a ring-shaped outcrop around the basin. Although they are nowhere really well exposed, they can be seen in the gorge of the River Ayr as well as in numerous road and railway cuttings (*Fig. 7.4*). We may envisage a desert landscape across which basaltic vents arose to heights of perhaps one or two hundred metres. Lavas flowing from these combined to produce a composite lava field in much the same way as molten wax from neighbouring candles might conjoin to form a single entity.

Lavas of the same age (*c.*290 Myr), still more indifferently exposed, outcrop again further south within the Southern Uplands terrane in the downfaulted desert basins of the Sanquhar and Thornhill areas. Apart from forming surface lavas, the basalt magmas also spread out beneath

Fig. 7.4 Columnar-jointed basalt lava within the Mauchline basin, Ayrshire.

the surface, within the Carboniferous sandstone and shale strata, forming extensive sills in the region. The notable south-facing cliffs of Benbeoch a few kilometres east of Dalmellington are formed from one such sill.

The Sound of Jura is the narrow stretch of water separating the islands of Islay and Jura: within it lies the small island of Glas Eilean, consisting of a stack of lavas with intercalated red sandstones. In the late Carboniferous or earliest Permian the sound was probably a narrow valley, bounded by a fault on the Jura side, that accumulated stones and sands from the rocky wildernesses on either side. In a modern desert it could be described as a wadi. The fault is presumed to have acted as a plane of weakness up which basalt magma arose to feed a small volcanic cone. Wind-blown sand, accumulated on the unweathered rubbly top of one of the lavas, gives testimony to the arid nature of the scene.

Highlands and Islands

In the Western Highlands, Inner Hebrides, Orkney and Caithness there are numerous small basaltic dykes that have been radiometrically dated as Permian or latest Carboniferous. Although the majority of these dykes are so small that they are very unlikely to have fed volcanoes, they are associated with basaltic plugs and diatremes up to about half a kilometre across, which almost certainly did reach the surface. However, their extrusive products have been eroded away other than at Glas Eilean (described in the section above). In the Western Highlands up to a dozen eroded diatremes occur from Kinlochleven to Applecross, with a particular concentration across the Great Glen Fault north of Fort William; finding these pipes usually requires dedicated search of the hillsides with the aid of a detailed geological map.

Dykes and sills of the 300 Myr event

At around 300 Myr something provoked a rather dramatic melting event that took place in the mantle. Precisely what caused this is, at the time of writing, a distinctly controversial question but it may have involved the uprise of a mantle plume (cf. Chapter 1) rising from the deep mantle to the base of the lithosphere, with the focus of activity well east of Scotland, somewhere beneath southern Scandinavia. Whatever the cause, the effect is beyond debate. N–S extension of the tectonic plate provoked the reactivation of E–W faulting in northern Britain and the ascent of magma into a major E–W trending dyke swarm. The volume of basaltic magma generated in this short-lived (a few million years?) event was certainly the largest for some hundreds of millions of years and was only matched and exceeded in the Palaeogene eruptions explored in chapter 5. It would appear that magmas fed laterally into the propagating dykes from the east so that the magma was not merely rising into the shallow crust but had a strong westerly movement.

The dyke swarm in Britain is approximately two hundred kilometres broad and although the more southerly dykes are in northern England, the majority transected southern and central Scotland as well as the southern part of the Highlands. The dykes range from a metre or so to several tens of metres in width and, from geophysical evidence, appear to thicken eastwards offshore into the North Sea. One beautiful outcrop occurs in Perthshire, east of the M90 at the Campsie Linn on the river Tay. This locality is of great interest in the history of the science since James Hutton correctly identified it as an intrusive dyke as early as the late 18th century.

Had the dykes attained surface level there can be no doubt that extensive fissure eruption of lavas would have ensued (so-called 'flood basalts'). There is, however, no evidence in Britain that this happened. In northern England and in the Midland Valley the magmas had to ascend through several kilometres of low-density sedimentary strata and, rather than making it to grass-roots level, the magmas intruded laterally between the beds of Carboniferous-age sandstones and shales to produce two extensive sills. One, the Whin Sill, was intruded across much of Northumberland, County Durham and Yorkshire. The present outcrops of this give rise to features such as the Farne Isles, the promontory on which Bamburgh Castle stands, High Force waterfall and a rocky escarpment which was exploited by Hadrian's construction workers in order to build his great wall. The Midland Valley Sill complex is the Scottish counterpart to the Whin Sill. Latest Carboniferous or Permian plate movements downfolded this into a broad syncline, the axis of which runs WSW–ENE from the vicinity of Cumbernauld towards East Fife. While most of the sill is hidden underground beneath a thick cover of Carboniferous sedimentary strata the peripheral rim is exposed inter-mittently over more than 160 km. The rim is studied with quarries as the dolerite of the sill is much prized as road metal and, in former times, was used for sets and kerb-stones. Some of its better known outcrops include the Lomond Hills of Fife, the foundations for the Forth bridges and the cliffs on which Stirling Castle stands. It is also to be seen in some deep road-cuts such as that on the A90 on the north side of the Forth road bridge.

We could regard this enigmatic 300 Myr event as one that merely possessed the potential for supplying volcanoes but which instead spent its force in feeding sub-surface intrusions. This view could, however, be errone-ous. Many of the dykes traverse the metamorphic rocks of the Highlands and indeed can be followed to the Outer Hebrides as on Barra. Magmas fed into these would not have been trapped within thick sedimentary sequences like those in northern England and the eastern Midland Valley. Consequently, while no evidence for accompanying volcanoes on the British side of the North Sea survives, the possibility cannot be totally excluded that in the Highlands and islands some fissural volcanism took place, the products of which have subsequently been totally removed by uplift and erosion. Lavas erupted in south Norway in the Skien area near Oslo have been attributed to this same magmatic event.

Late Carboniferous volcanoes

Some tens of millions of years earlier than the volcanism in the Mauchline area of Ayrshire, activity of a similar type occurred between 320 and 290 Myr. The remains of these volcanoes are concentrated around the south side of the Clyde Estuary and, in the east, around the Forth Estuary, partic-ularly in Fife. At this time the British region still lay in tropical latitudes but the various components of Pangaea had yet to be fully assembled. Plate motion had not yet closed the seaways so that, in contrast to the arid times that were to follow, the moist sea winds still blew. The resultant rainfall ensured that the landscape was lushly vegetated and was sufficient to supply rivers that flowed south from the decaying Highland mountains towards the Midland Valley. In these times the Highlands were the extensively worn-down stumps of what once had been a majestic mountain range thrown up in the Caledonian Orogeny, about one hundred million years earlier.

The rivers deposited their loads of silt and sand in extensive subsiding deltas. These delta lowlands would have looked rather similar to those of the present-day tropical deltas of, for example, the Irrawaddy, Niger or Amazon rivers, supporting extensive swamp forests. The vegetation, however, was utterly different from that of modern tropical forests, being composed of giant ancestors of today's horsetails and clubmosses. It was a period when the atmosphere had exceptionally high oxygen but low carbon dioxide contents compared with those of today. Within these primeval forests giant insects crawled or flew while salamander-like amphibians were kings of the jungle. The great thicknesses of dead and decaying plant debris derived from these swamp forests that accumulated on the slowly subsiding deltas formed the coal-seams that give the Carboniferous its name. The subsidence of the deltas was episodic. Sinking allowed inundation and the drowning of the forests, but these were repeatedly restored as further deposition of river sediment allowed shoaling and re-establishment of the land plants.

Between 320 and 300 Myr in these moist tropical environments there were periodic eruptions from a multitude of small volcanoes in the eastern sector of the Scottish Midland Valley. In Fife alone there are over one hundred vents filled by fragmental material and basaltic intrusions. Those magma batches which made it all the way to the surface encountered groundwater or water-saturated sediments and violent episodes (boiler explosions!) resulted from the generation of high-temperature steam. *Fig. 7.5* shows a recent tuff-ring of this type formed in equatorial conditions on one of the Comores Islands between East Africa and Madagascar. The late Carboniferous volcanoes of East Fife would have presented a similar appearance.

As is so generally the case with these old volcanic structures, subsequent uplift of the land by deep-seated processes was accompanied by erosion from the top down, and what we see in Fife are cross-sections of the roots of these diatremes at levels of two or three kilometres down. These roots normally have the form of carrot-shaped bodies, typically occupied by pyroclastic rocks, widening upwards as shown in *Fig. 7.3*. Intrusions of the basalt magma into and around the diatremes are commonly dykes up to a few metres wide, although irregular intrusions are common in the poorly stratified fragmental rocks. One particularly prominent dyke of this diatreme association can be seen at low tide on the south coast of Fife, cutting the rocks of the Ardross vent, about two and a half kilometres NE of Fife.

Fig. 7.5 Air-photograph of a tuff-ring on the island of Grande Comore in the Mozambique Channel, about 12° south of the Equator.

Lavas associated with these eruptions would have been of fairly trivial volume and are scarcely ever preserved. While the inland vents are poorly exposed there are some beautiful exposures in coastal sections including those at Kincraig, Elie Harbour and Elie Ness on the south coast of East Fife. A typical vent contains well bedded strata largely composed of broken fragments of country-rock shale, sandstone and basalt. Along with these are better rounded masses, up to a metre or so across, whose form indicates that they were ejected as molten magma that assumed their sub-spherical form through surface tension forces while airborne. The bedding in the pyroclastic products is due to alternation of coarser and finer strata which probably reflects both the pulsatory nature of the explosive eruptions as well as varying wind strengths that helped to sort denser from lighter particles falling from the erupting column of gas and ejected solid and molten particles. The strata on the outer flanks of the tuff-rings had outward dips while those on the inner side of the craters dipped inwards (*Fig. 7.3*). However, when the explosive eruptions ended, much or all of the surface structure collapsed back into the vent and the attitude of the bedding planes that we now see is typically steep and inward dipping as a result. *Fig. 7.6* shows poorly-stratified basaltic pyroclasts from one of these diatremes, at Elie Harbour, Fife. Black, carbonised remains of plant stems, commonly encountered within the bedded ashes, provide vivid testimony to the luxurious growth of vegetation on and around these volcanoes.

Fig. 7.6 Basaltic bombs and broken fragments showing crude stratification. Elie Harbour vent, Fife. The graduated rod is half-a-metre long.

The geological evidence in Fife thus presents a scene of tropical forested deltas with scattered volcanic eminences providing much of the higher ground. It is improbable that any two of the volcanoes ever erupted precisely at the same time, and quiescent intervals of hundreds of thousands of years may have elapsed between short-lived episodes of explosive excitement. There is evidence in the coal-seams indicating that the forests were periodically destroyed by fire. In so highly oxygenated an atmosphere green vegetation would have been almost as much at fire risk as dead wood. Whereas most such fires are likely to have been triggered by lightning strikes, the occasional volcanic eruption would also have constituted a major fire hazard in these exceptional conditions.

As with most of the later Carboniferous–Permian volcanism described earlier (but excluding the anomalous 300 Myr dyke and sill forming event), the magmas are believed to have been the products of small-scale melting at depths of 70–90 km. The resulting basaltic magmas had relatively high concentrations of carbonate in solution which, during ascent and depressurisation, decomposed to produce fluid carbon dioxide. The appearance of this would have lowered the density progressively as the fluid-filled vesicles expanded as pressure decreased with the consequence that the magmas accelerated upwards, possibly to rates measurable in terms of metres per second. These dramatically fast rising magmas not only had the energy to wrench off pieces of rock from their side walls but were also able to flush them up to the surface. These pieces, referred to as 'xenoliths' (literally, 'foreign stones') include fragments of mantle peridotite carried up from depths of over 30 km. These are high-density materials of up to cauliflower size and their presence provides dramatic evidence for the very energetic rise of the magma which brought them up. Often the xenoliths were surrounded by a thin coating of molten basalt. One may picture them as rather like toffee-apples where the apple is replaced by the peridotite and the toffee by an adhering layer of the basaltic lava. Additionally the vents and associated intrusions contain (rare) fragments of rock from the lower crust. That these are rarer than the mantle xenoliths stems from the fact that the crustal rocks are much more readily melted by and dissolved in the host magma. For any piece to have survived the journey requires that it started large, was carried fast and finally cooled quickly otherwise, like a sugar-lump in hot coffee, it would have vanished from sight. The deep-source xenoliths in these late Palaeozoic basaltic volcanoes are important since they provide the only direct and tangible evidence (as opposed to indirect evidence from chemistry and physics) of the nature of the rock formations present at depths from *c.*10–70 km below Scotland. For comparison it is worth remembering that the deepest drill-holes and mines penetrate less than 5 km!

At around 315 Myr and generally contemporaneous with the activity in East Fife, basaltic eruptions occurred SSW of where Glasgow now stands. These so-called Passage Group lavas produced a composite lava field some 170 m thick in the Saltcoats–Kilmarnock–Troon area. These underwent extensive penecontemporaneous weathering in the hot humid climate, with pervasive leaching and formation of red lateritic rocks. Much of the lateritisation would have been associated with the physical and chemical activity of plant roots exploiting the newly-formed eruptive products. Some of the weathered volcanic rocks have concentrations of iron high enough for them to have been worked in the past as low-grade iron ores. Very little is known of the volcanoes from which they were erupted. While some landforms would have been of the open 'tuff-ring' type formed by the explosive release of steam through the interaction of magma and surface or near-surface waters, and composed largely of pyroclastic materials, others, from which the bulk of the lavas were emitted, would have been shield volcanoes with subdued profiles.

Chapter 8

Post-Caledonian Relaxation: the Lower Carboniferous Volcanoes

This chapter continues the backward look through the late Palaeozoic volcanic history of Scotland. Starting at around 320 Myr it traces the tale back to around 359 Myr at the very beginning of the Carboniferous. During the first twenty-five million years of the Carboniferous an ocean closure was proceeding hundreds of kilometres south of Scotland as the tectonic plate bearing the great continent of Gondwana (incorporating what is now Africa, India, South America and Antarctica) converged on the Laurasian continent (within which lay Scotland) to the north (*Fig. 7.1*). The ensuing continent to continent collision brought about a great mountain forming event known as the Variscan Orogeny, whose traces are now found across the southern parts of Ireland, Wales and England as well as in Spain, France and Central Europe. Although these collisional happenings occurred far to the south of Scotland, the shuffling of continental blocks in northern Britain, involving lateral movements with some degree of 'pull-apart', lowered the pressure on the underlying mantle to the extent of causing partial melting with the genesis of basaltic magmas. Like that described in the previous chapter, this early Carboniferous volcanism can all be described as 'intra-plate'.

It was as a response to these releases of pressure that, after some 40 million years of Devonian tranquillity, volcanoes became active in the south of Scotland around 350 Myr; from then eruptions took place, on and off, across southern, central, western and far north-eastern Scotland right through the Carboniferous and, as we have seen (Chapter 7), on into the Permian. Consequently any division of the volcanism in the early Carboniferous (up to about 320 Myr) from the subsequent later Carboniferous and Permian activity is wholly artificial. However, the fact that the later activity tended to be on a much smaller scale than in the early Carboniferous provides the reason for dealing with the latter separately in this chapter.

The story is resumed in the interval 320–330 Myr, before the 'Passage Group' lavas were erupted, when the Midland Valley was still experiencing

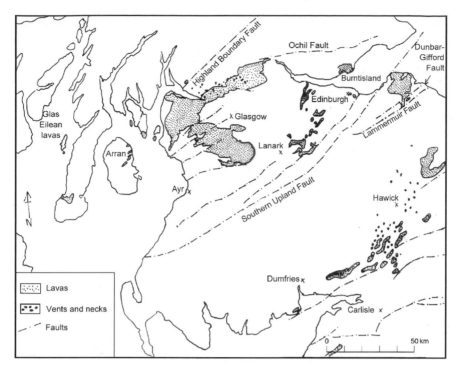

Fig. 8.1 Map of South and Central Scotland showing distribution of early Carboniferous volcanic rocks. *(After W. A. Read et al., 2002)*

humid equatorial conditions. First we shall consider some volcanoes in West Lothian which represent the waning stages of activity in the early Carboniferous (*c*.330–340 Myr) that was on a vastly greater scale than anything the late Carboniferous and Permian had to offer. In the Bathgate Hills of Midlothian, extending southwards from Bo'ness on the Firth of Forth towards the town of Bathgate, basaltic lavas and ash beds outcrop on the eastern margin of the Central Coalfield. These strata, with a thickness totalling around 600 m, date from around 340 Myr, which is more or less in the middle of Carboniferous times. The Bathgate volcanic rocks piled up to a slightly elevated ridge which, for most of the time of their accumulation, was a little above sea-level. Although high and dry for the most part, it was occasionally submerged beneath the seas during episodic sea-level highs.

Global sea-levels at this time oscillated in response to growth or shrinkage of polar ice sheets and, in the warmer intervals, shallow seas encroached across the lowlands of the Midland Valley. During some of these marine inundations, the discharge of sand and silt by rivers draining from the

Fig. 8.2 Saba a volcanic island in the Zubair group, in the southern Red Sea. Coral reefs have grown on and around basaltic cinder cones.

Highlands into the seas was barely perceptible and the resultant clear warm waters provided ideal habitats for the growth of coral. Basaltic volcanoes, however, still made sporadic appearances, growing from the sea floor to form islands around which, during prolonged periods of repose, coral reefs grew. Some of these reefs have survived, fossilised as limestones. There are quite a few locations in the British Isles where this combination of black volcanic rock and pale coral limestone can be found. These locations range from south-west Ireland to Derbyshire and the Isle of Man but there is a Scottish example in the Bathgate Hills on the southern side of the River Forth. *Fig. 8.2* shows a basaltic volcano surrounded by coral reefs at the southern end of the Red Sea, and these early Carboniferous environments would have looked rather similar.

The East Kirkton: a lake in a volcanic landscape

Apart from the occasional marine inundations, the Bathgate ridge generally stood high and dry above sea-level. Its component volcanoes were little ones, unlikely to have risen more than one or two hundred metres high. It was through a remarkable combination of circumstances that evidence of the ecosystem that this hilly landscape supported was fossilised and preserved. As has been pointed out earlier, lakes may form in volcanic terranes where rainwaters pond within craters or calderas or as a result of rivers becoming dammed (for example by lava flows or debris flows slumping from volcano flanks). In Chapter 5 a famous fossil flora (and accompanying fauna) was described from lake deposits within the Palaeogene lavas at Ardtun on Mull. In the Bathgate Hills we now encounter another outstanding fossil locality, also involving a palaeo-lake surrounded by quiescent volcanoes, but this time some 280 million years older than those of Ardtun. Its deposits, exposed in the course of quarrying at East Kirkton, near Bathgate, were found to contain a globally unique collection of plant and animal fossils. As the realisation dawned that these rocks provide a remarkable opportunity to study the ecology of a long-lost Carboniferous world, they became subjected to intense scrutiny by a host of palaeontologists and other specialists.

From this research an astoundingly detailed picture slowly emerged of the ecosystem that existed in and around the East Kirkton lake. All lakes are ephemeral features and this one probably existed for only a few thousand years. While its size and shape are unknown, it appears to have been shallow and subject to occasional drying-out. The deposits that accumulated within it, to a thickness of 11 m, were of layers of oily mud and basaltic ashes, interstratified with fresh-water limestones and siliceous (chert) horizons. The mud and volcanic debris represent materials washed into the lake while the limestones and cherts were precipitated directly from the lake waters. Despite the fact that a cornucopia of plant and animal fossils has been recovered from the deposits, the lake itself appears to have been home to few living organisms, and most of the fossil organisms in its sediments were swept into it after their deaths.

It has been established that hot springs fed into the lake and that the waters themselves were either periodically too hot (up to 60° C) or too chemically polluted to support any but highly specialised life forms. From the detritus that accumulated on the lake floor, however, a vivid picture of the enchanted forests that surrounded it can be visualised (*Fig 8.3*). Dozens of plant species grew around the lake and the fronds of many are beautifully preserved along the bedding planes of the sediments. The principal forest trees, with trunks up to 50 cm across, were ancestral forms of modern tree-ferns and conifers (pteridosperms and gymnosperms). Many thin coaly horizons in the sediments are of a type known as vitrain, which is fossilised wood ash (signifying forest fires). Forest fires were clearly of common occurrence since the oxygen content of the Carboniferous atmosphere was higher, and it has been supposed that many of the animal remains are those of creatures that entered the lake to escape the flames. The fossils indicate that the forest supported a very prolific fauna. Among the invertebrates were myriapods (relatives of millipedes and centipedes),

Fig.8.3 Palaeoecology of East Kirkton lake. Reproduced by permission of the Royal Society of Edinburgh and E.N.K. Clarkson, A.R. Miller and M.I. Coates in *Transactions of the Royal Society Edinburgh: Earth Sciences*, volume 84 (1994, for 1993), pp 417–425.

arachnids (mites and harvestmen, true spiders having not yet evolved) and members of the extinct group of arachnid-like creatures called eurypterids, up to a metre or more long. The eurypterids probably lived around the lake-shores, feeding on shrimps in lakeside pools or shallows. We may suspect that primitive insects were also present – although no fossils have yet been reported. Flying insects had not yet evolved and the forests are likely to have been rather silent as well as devoid of the colours of flowers. The remains of scorpions, up to 70 cm long, are among the fossils and were probably the principal predators in the forests.

The plants and invertebrate animal fossils from these deposits would have been quite sufficient in themselves to make East Kirkton a world-famous locality, but even greater excitement came from the discovery of its vertebrate fossils. Fish fossils do occur at a few horizons, including a small, dogfish-like shark. These suggest that at times the lake, which over most of its history was inimitable to fish, occasionally had some connection with fresher and probably better oxygenated waters. However, the most remarkable finds were of the remains of several species of tetrapod (literally, four-footed) vertebrates. A number of specimens of salamander-like amphibia have been described (*Fig. 8.3*) together with an unusual amphibian species which was elongate and legless and which had probably adopted a lifestyle comparable to that of modern snakes. True reptiles did not appear until the Permian and Triassic, but the tetrapods from East Kirkton appear to have been well on their way towards acquiring reptilian characteristics.

The reptiles, birds and mammals of recent times are grouped together as amniotic animals, i.e. those whose embryos are protected by an amniotic membrane which permits production of offspring either from eggs laid on land or within the body of the mother. In contrast, fish and amphibians require water in which to lay their eggs and their young grow through a larval form. Some of the East Kirkton tetrapods are deduced to have been 'stem-amniotes', or almost-but-not-quite amniotes! They certainly predate by 40 million years what had hitherto been the earliest known fossil amniote, from Nova Scotia. One of the East Kirkton species, *Westlothiana lizziae,* was actually thought to be a lizard-like reptile when first described. However, although 'Lizzy', as it came to be known, was subsequently shown to lack true reptilian features it and its companion species may be regarded as intermediary 'missing links' between the amphibian and reptilian stages of evolution.

These amphibians were slowly adapting to a mode of life where water was no longer critical for their juvenile development and were evolving to fully terrestrial reptiles. We must imagine that at about this time, our distant ancestral grandmothers had mastered the knack of raising their offspring on land, without the necessity of an aqueous nursery. Clearly such fossils must be treated with respect, not only owing to their scientific value but because they are potentially ancestral to us. It must also be a matter of some pride that this earliest semi-terrestrial fossil is Scottish!

The perfect preservation of so many plant and animal fossils at East Kirkton was due to a mineralisation process attributable to the percolation of warm solutions through the lake muds. These hydrothermal solutions were the result of the immediately preceding magmatism. Although volcanic terranes do not usually provide happy hunting grounds for palaeontologists, East Kirkton, because of a remarkable combination of circumstances, is of critical importance for the light it casts on terrestrial ecosystems in the mid-Palaeozoic Era. In the following chapter we shall meet another remarkable locality where immaculate details in the fossils are retained thanks to volcanic hot springs.

The Clyde Plateau volcanoes

The Bathgate Hills activity represented the waning end stages of a far more vigorous stage of volcanism which occurred in the early part of the Carboniferous from 342 to 335 Myr. In this period more than 90% of the total volume of Carboniferous–Permian lavas were erupted. The bulk of these were spilled in the Midland Valley and Scotland had to wait for the Palaeogene volcanoes of the Hebrides, some 270 million years later, before comparable quantities of lava were erupted again. The main focus of action was the western part of the Midland Valley, where we see the products outcropping in the great chevron-shaped belt of hills that encloses the Glasgow basin on its northern, western and south-western sides. This belt of hills is collectively referred to as the Clyde Plateau. The lavas accumulated to form a volcanic field some 75 km E–W and about 50 km N–S, with a maximum thickness of close on 1,000 m. In later Carboniferous times the whole lot was deformed into a broad downfold (syncline), the axis of which dips gently east-south-east, trending from the vicinity of Greenock through Glasgow towards Motherwell and beyond. The upraised edges of this V-shaped syncline, exposing the volcanic rocks, form the high ground, incorporating the Touch, Gargunnock, Campsie and Kilsyth Hills and the belt of moorland stretching from Greenock south-east to the Eaglesham and Strathaven districts. The lavas in the central part of this downfold are hidden beneath the younger sandstones and shales of the Central Coalfield so that in, for example, central Glasgow, the lavas are several hundred metres below the present surface.

In these relatively early Carboniferous times the Clyde Plateau would have been a region of elevated ground rising above the surrounding sandy fluviatile plains and marshes that lay little above sea-level. The Midland Valley lay well to the south of the Equator. Although the lowlands would have been green and well vegetated, the forests would not have attained anything like the height and luxuriance that they achieved fifteen to twenty million years later when the great coal swamp-forests of the late Carboniferous were established. The more sparsely forested and better drained volcanic uplands would have presented an undulating topography crowned by the larger shield volcanoes and occasional steeper cones of composite volcanoes. The ruins of some of the bigger composite volcanoes are now represented by Meikle Binn in the Campsies (*Fig. 8.4*) and by the intrusive/extrusive complexes of Irish Law and Misty Law in the moorlands south-east of Greenock. Although big by Carboniferous standards they were much smaller than the giant Palaeogene edifices discussed in

Fig. 8.4 Meikle Beinn in the Campsie Hills. An eroded remnant of one of the larger Clyde Plateau central-type volcanoes.

Chapter 5 and also small in relation to some of their predecessors that form the subject of the next chapter.

The whole Clyde Plateau volcanic field was a composite structure composed of mainly basaltic lavas erupted from a multitude of relatively small scoria cones, lava shields and, occasionally, fissure volcanoes. We may picture the lavas as generally having been of the rough and jagged topped *aa* type (*Fig. 4.2*) rather than the more mobile and smoother surfaced *pahoehoe* kind (*Fig. 4.3*). In the earliest phases of the Clyde Plateau volcanism the basaltic magmas encountered surface waters and waterlogged strata and explosive steam-propelled eruptions ensued giving rise to basaltic ash layers. However, as more and more magma welled up from the depths and formed near-surface intrusions, the ground became elevated above the water table and relatively passive outwelling of basaltic lavas became the norm, with explosive activity becoming rarer.

On a clear day the view south from the road between Stirling and Doune affords a stunning view of the north-facing lava escarpment of the Gargunnock Hills (*Fig. 8.5*). Here the stepped profile of the escarpment is due to the more readily weathered flow-tops alternating with the tougher, more resistant, flow interiors that provide the steeper faces. The lavas of the Campsie and Touch Hills further to the south-west appear to have been

Fig. 8.5 A view of the Gargunnock Hills showing the regular terracing due to preferential erosion of the weaker rocks composing the tops of lava flows.

erupted from a large number of volcanoes sited along a continuation of an ENE–WSW lineament extending some 45 km from Dumbarton in the WSW towards Stirling in the ENE. The sites of these volcanoes are now marked by plugs or vents filled with fragmental materials. *Fig. 8.6* shows an illustrated cross-section of the volcano that once overlay Dumbarton Rock. An excellent place to examine the lava succession is the Campsie Glen, where a succession of waterfalls marks the harder flow interiors, with the intervening flatter stretches denoting the contact zones between one lava and the next.

The Stirling–Dumbarton lineament probably marks a fault plane at depth, activated in the early Carboniferous and allowing the magmas easy access to the surface. In latest Carboniferous–early Permian times further reactivation of this fault was responsible for what is now the great south-facing escarpment of the Ochil Hills north-east of Stirling. The eruptions along the lineament probably came from a complex array of shield volcanoes, although the great lateral continuity of flows seen in the Gargunnock and Touch Hills suggests that fissure volcanoes played their part. The fault (or flexure) zone down-dropped on its south-eastern side giving a south-east facing escarpment with eruptions occurring along its base. The lavas from these are inferred to have flowed southwards, away from the flexure and the higher ground to the north. This is precisely the opposite of what we now

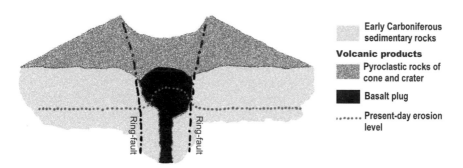

Fig. 8.6 Diagrammatic illustration of the growth of the Dumbarton Rock volcano. *(After J.G. MacDonald, in* Carboniferous and Permian Igneous Rocks of Great Britain, *Geological Conservation Review Series, ed. Stephenson et al., 2003)*

observe in the topography where the lavas form the high ground, with their escarpment facing northwards across the broad valley towards the edge of the Grampian Highlands. It would appear that we have here a fine example of inverted topography as discussed in Chapter 5, where what was once low ground became filled with erosion-resistant lavas while erosion preferentially removed the bounding, less resistant, sedimentary rocks, turning what was formerly high ground into low ground and vice versa.

Although there is no doubt that the Clyde Plateau lavas extend beneath the surface, eastwards below Glasgow and the Central Coalfield, geophysical investigations indicate that the lava succession wedges out rapidly in this direction. Likewise, the available evidence shows that the lavas never spread very much further west than their present outcrops. Thin outliers of these successions form the island of Little Cumbrae and the southern point of Bute but these appear to be at about the most extreme westerly limits reached by the lavas. Although evidence from Ben Bowie, just west of Loch Lomond, shows that locally the lavas did reach just north of the Highland Boundary Fault, this was exceptional and the likelihood is that the lavas of the northern limb never travelled much further north than the present Touch–Gargunnock–Campsie Hills escarpment. Possibly there was higher ground to the north that prevented their further encroachment in this direction. Further instances where the Clyde Plateau lava distribution may have been controlled by pre-existing escarpments blocking their passage are seen in the region between Greenock and Strathaven. The WSW–ENE-trending Dusk Water and Inchgotrick faults were probably seismically active in the early Carboniferous, giving rise to north-facing escarpments that acted as barriers to the spread of the lavas.

Heads of Ayr

Much of the Clyde Plateau is poorly exposed but a remarkably fine set of coastal exposures across an isolated volcanic vent, active at the same time as the Clyde Plateau eruptions, can be seen at the Heads of Ayr. This headland is situated on the coast of the Firth of Clyde, about 5 km SW of Ayr. The exposures are in steep cliff sections but are also beautifully seen in the wave-cut platform revealed at low tide. The remains of this volcano have much in common with those described in the previous chapter from the later Carboniferous examples on the Fife coast. Like the latter, it grew explosively through the interaction of surface or near-surface waters with ascending magma. A tuff-ring built up on the surface was composed of fragmental material, principally volcanic but involving some blown-out fragments of the sedimentary wall rock. Like the Fife examples, what we see now is a section, well below the original land surface, displaying materials that have collapsed back down the vent from which they were blown, as a result of late-stage subsidence within sub-cylindrical ring faults. The structure as we see it has a diameter of about 850 m, but the diameter of the former tuff-ring may have been two or three times greater. The nature of the pyroclastic deposits suggests that they were in part air-fall from eruptive clouds and partly deposits from surges driven by expanding gases moving laterally at high velocities outwards from the explosive focus. Among the bedded air-fall rocks are sizeable rounded 'bombs' that would have been semi-molten masses of newly formed magma that impacted in and around the volcanic crater.

Compositionally the Heads of Ayr magma differed from those of the Clyde Plateau in general, in being of a type resulting from a smaller degree of mantle melting than the latter and rising from greater depths. It is of a silica-poor type of basalt essentially identical to those of East Fife and is of interest in being one of the earliest examples of this variety seen in Britain or, indeed, globally. Like those of the much younger Fife volcanoes (Chapter 7), these deep-origin silica-poor basalt magmas are inferred to have ascended at high speed, ripping off and carrying up pieces of the lithospheric mantle and crustal wall rocks. Consequently, as in Fife, pieces of peridotite and other ultramafic mantle components, incorporated by the magma at depths of up to 60 km, were shot up to the surface and blown out along with all the other volcanic debris. Heads of Ayr is noteworthy in being among the very few early Carboniferous localities where this phenomenon is exhibited.

Southern Kintyre

Almost all of the early Carboniferous volcanism took place in the Midland Valley or in the Southern Uplands and Borders. Some, however, occurred north of the Highland Boundary Fault, and we see the products in the southern part of the Kintyre Peninsula. Across the North Channel, close to Ballycastle in Antrim, we find a similar situation. The principal evidence for volcanoes of this age in the Highlands comes from a belt inland from Machrihanish in the form of a pile of lavas 400 m thick. The southern Kintyre volcanoes appear to have been operational at the same time as those of the Clyde Plateau and, no doubt, arose from the same causes. Despite their synchronicity the Kintyre volcanic field is believed to have been geographically distinct from, and never contiguous with, the Clyde Plateau lavas. The Machrihanish succession acquired a dip towards the SE as a result of later earth movements. Like the Clyde Plateau succession, it appears to be a complex sequence composed of overlapping flows from a number of different vents. There is some evidence that faulting accompanied the volcanism and that newly formed fault escarpments acted as barriers and constraints to subsequent lava flows. As outline above, similar phenomena are thought to have attended the build-up of the Clyde Plateau succession.

Compositionally the south Kintyre lavas have a number of idiosyncratic features that distinguish them from their Clyde Plateau contemporaries. Firstly, some lavas near the base of the succession are unique among the Scottish Carboniferous–Permian basalts in that their chemistry implies that they were produced by unusually extensive (and relatively shallow-depth) melting in the mantle beneath. We really have to look to some of the Hebridean Palaeogene basalts to find equivalents. Secondly, the lavas can be subdivided into a lower set, which are essentially mafic (basaltic), and an upper part characterised by more evolved compositions, grading through from basalt to trachyte. Trachyte, produced at a late stage in the activity, caps the hill of Skerry Fad. Trachyte lavas, typically 'thick', viscous and slow-moving, rarely flow far from their eruptive vent, and Skerry Fad may represent a stodgy capping ('lava dome') overlying its feeder pipe. That the more evolved magmas were erupted later in the activity carries the implication that their parental (primitive) magmas had very slow overall rates of ascent through the lithosphere (probably getting held up at the mantle–crust boundary) and had greater times to crystallise at depth, so modifying their chemistry to greater or lesser degrees. It is commonly observed that, at the start of magmatism, the magmas rise fast and that, with the passage of time, energy wanes and fractional crystallisation is able to proceed to a greater extent.

West and central Fife

In the vicinity of Burntisland and Kirkcaldy, to the east of the Midland Valley and north of the Firth of Forth, volcanoes erupted at about the same time as those of the early Bathgate succession and at the very end of the Clyde Plateau activity. These built up a lava pile that now forms the high ground above Kinghorn. The basalt lava escarpment here played an important role in Scottish history when King Alexander III, riding in haste to join his young bride, fell over it to his death in 1286.

Edinburgh and Arthur's Seat

Of all the ancient volcanic relics in Scotland the most celebrated are those within the confines of the city of Edinburgh that form the upstanding features of the Castle Rock, Calton Hill and Arthur's Seat (*Fig. 8.7*). Since these attract the attentions of huge numbers of tourists they will be described in some detail. But first, let us recall the overall situation in which these rocks first formed. For some tens of millions of years during the closing stages of the Devonian and the early part of the succeeding Carboniferous Period, sands and muds had been accumulating in the Edinburgh region, deposited mainly by sluggish rivers. What is now the Firth of Forth, Fife and the Lothians would have been tropical lowland close to sea-level. It was in such an environment that, at very much the same time as the Clyde Plateau volcanoes came into play and instigated by the same earth movements, volcanism broke out at around 340 Myr.

The magmas reached levels of neutral buoyancy as they approached the surface, particularly in the low-density sedimentary strata, where they spread out laterally as sills. Some batches, however, clearly reached surface level and erupted as small basaltic volcanoes. Initial release of gas, largely steam from heated ground-waters, drilled sub-cylindrical conduits which were followed and enlarged by rising magma. Edinburgh Castle stands upon a sub-cylindrical mass of basalt that almost certainly represents the infilled conduit beneath a former volcano. The basalt is ovoid in plan, about 300 x 200 m across, elongated NW–SE, and cuts across earlier Carboniferous sandstones. The Castle Rock is likely to have been surmounted (probably not over one or two hundred metres above the present top) by a cinder cone itself a few hundred metres high, with an external diameter of about 2 km. In other words the Castle Rock volcano would have covered most of the area of the modern city.

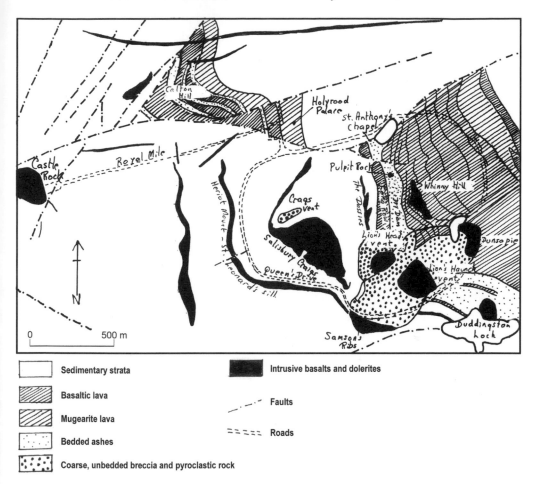

Sedimentary strata

Basaltic lava

Mugearite lava

Bedded ashes

Coarse, unbedded breccia and pyroclastic rock

Intrusive basalts and dolerites

Faults

Roads

Fig. 8.7 Geological map of Edinburgh and a cross-section of Holyrood Park. *(After Land and Cheeney, 2000, Carboniferous and Permian Igneous Rocks of Great Britain, in Geological Conservation Review Series, ed. Stephenson et al., 2003 and Mitchell and Mykura, 1962)*

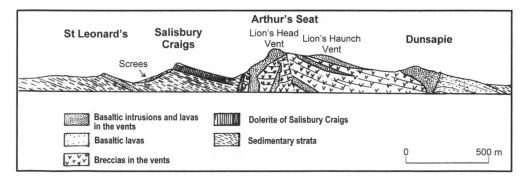

Basaltic intrusions and lavas in the vents

Basaltic lavas

Breccias in the vents

Dolerite of Salisbury Craigs

Sedimentary strata

The volcanic rocks forming the high ground in Holyrood Park (including Arthur's Seat) and Calton Hill were produced at much the same time. The two or three thick lava flows that make up Craiglockhart in the southern part of the city were formerly thought to have been distinctly older, but subsequent investigations suggest that these too were essentially contemporary. In composition and origin all these rocks are closely allied to the much more extensive lavas of the Clyde Plateau, although they were most probably quite separate from the latter.

Calton Hill and much of the Holyrood high ground is composed of lavas and ash layers with numerous intrusions and vents filled by fragmental deposits. The volcanic succession is some 200 m thick on Calton Hill, thickening to more than twice as much on Whinny Hill on the eastern side of Holyrood Park. Although the strata would have been not far off horizontal, with dips of less than 10° at the time they formed, some tens of millions of years later they were deformed as part of the deep down-fold (syncline) of the Midlothian Coalfield. Consequently, lying within the western limb of this syncline, the rock layers in Calton Hill and in Holyrood now dip eastwards into it, at angles of 20° or more. The Holyrood and Calton Hill volcanic strata are all part of the same sequence. Thus Calton Hill shows a number of lavas with subordinate ash layers, all dipping east as described above.

Glaciers flowing from the west during the past one million years have sculpted the topography of today's Edinburgh. Among the most obvious results of their work is the classic 'crag-and-tail' topography exhibited by Castle Rock and the gentler gradients of the 'Royal Mile' ridge to the east, the steep west-facing escarpments of the Salisbury Crags Sill and Whinny Hill lavas and their eastern dip-slopes as well as the analogous dip and scarp forms of Calton Hill. In each case the steep west-facing escarpments are due to igneous rocks that presented more resistance to the glaciers, which consequently differentially eroded the softer sediments and fault zones.

In Holyrood Park and Calton Hill, volcanism commenced with explosive activity yielding an ash layer (currently unexposed) immediately beneath the first lava. There have been different opinions concerning the number of lavas in the Whinny Hill succession in the northern part of Holyrood Park, with estimates varying between nineteen and thirteen. Lava 1, which is approximately 30 m thick, forms the prominent cliff referred to as the Long Row. This lava appears to be missing from the Calton Hill suite but outcrops again south of the Lion's Haunch Vent in the vicinity of Duddingston Loch. Compositional similarities between the Castle Rock basalt and lava 1 led George Black, who worked extensively on the geology of the park in the 1950s, to suggest that it erupted from the Castle Rock volcano and flowed eastwards. While this is a reasonable inference it has not been verified by later workers.

Above lava 1 is a 30 m thick sequence of ashy mudstones and limestones containing plant fragments, thought to have formed in lagoonal waters close to sea-level. Being easily eroded this sequence has weathered out to form the terrace feature known locally as the Dry Dam. It is overlain by an ash layer signifying resurgence of volcanism and then by a severely altered basalt flow (lava 2) about 8 m thick. This in turn was succeeded by the 7 m thick 'Upper Ash of the Dry Dam'. That this contains fragments of plant and fish fossils suggests that it was still very close to sea-level. However, from the Upper Ash onwards volcanism was continuous for a considerable while and the succession is wholly composed of lavas and pyroclastic rocks.

The third lava has a very limited outcrop confined to a restricted zone extending a little over 200 m south from the ruined St Anthony's Chapel. It shows quite good columnar jointing in its lower parts but abrupt changes of attitude in the columns, from vertical at its base to sub-horizontal near the chapel, testify to a complex cooling history. The upper part of the lava, however, is distinctly blocky with irregular jointing. The blocks vary in their degree of vesicularity and some were frothy with abundant gas bubbles (vesicles). We are almost certainly looking at what were partly molten blobs that accumulated on the surface of the lava, very close to an erupting vent. According to George Black lava 3 erupted from a volcano sited above the basaltic knoll called the Pulpit Rock, a short distance SSE of the chapel. The Pulpit Rock, on the western flank of Whinny Hill, appears to be a basalt neck which would have been overlain by a volcano, not many metres above the present top of the rock. We may imagine a clinkery cone of blocks and spatter from lava fountains and a single lava that flowed northwards towards the St Anthony's Chapel area.

The next lava (lava 4) is about 8 m thick and outcrops to the south of the Pulpit Rock, overlying a layer of basaltic ashes. It has well developed columnar jointing and forms the notable escarpment above the eastern flank of the Dry Dam. A succession of lavas lies above with lavas 5, 6 and 7 forming distinct west-facing escarpments around the top of Whinny Hill. The uppermost six or so lavas, which make much less distinctive landforms on the eastern flanks of the park, dip down towards Portobello. All of this succession is transected to the south by two successive volcanic vents, the Lion's Head and Lion's Haunch Vents. The Lion's Head Vent is the smaller and older. Before its intersection by the Lion's Haunch Vent it would have been approximately circular in plan with a diameter of some 300 m. It is filled with broken pieces of volcanic and (scarcer) sedimentary rock, up to about 10 cm across. All this fragmental material (or volcanic breccia), which shows a crude bedding dipping centrally, has been penetrated by a number of basaltic intrusions. A complicated bundle of basaltic dykes in the lower exposures coalesces upwards to form a coherent mass in the centre of the vent, now forming the summit of Arthur's Seat. Quite well developed fine-scale columnar jointing is seen around the summit. It is reasonable to conclude that the Lion's Head breccia formed as scree or talus of fallen rocks within a crater that had been excavated by explosive (steam driven?) eruptions, and that when this violent phase had ended basalt magma found its way up through the jumbled mass of broken rock and may well have produced a lava lake within the crater. George Black suggested, on rather slender evidence, that lavas 2 and 4 had flowed northwards from the Lion's Head Vent. Lava 3, as pointed out above, was probably emitted from the Pulpit Rock, which may have represented a smaller basaltic cone parasitic on the northern flanks of the Lion's Head Volcano.

The younger, Lion's Haunch Vent has an oval plan, extending approximately 1,200 m NE–SW by 500 m NW–SE. The vent includes Dunsapie Hill in its north-eastern extremity as well as the basaltic intrusion of Samson's Ribs at its south-west. The volcanic breccia which occupies much of the Lion's Haunch Vent is, like that of the Lion's Head Vent, crudely layered with the bedding dipping towards the interior of the vent. It tends to be distinctly coarser than that of the Lion's Head, with basaltic chunks up to 2 m across, accompanied by some pieces of sandstone, mudstone and limestone. The contents differ from those of the earlier vent in that they include fragments of the higher (and more evolved) Whinny Hill lavas. The breccia is most conveniently examined in the roadsides along the northern margin of the Queen's Drive between Samson's Ribs and the Lion's Haunch. As in the Lion's Head Vent the breccias were almost certainly produced during periods of explosive discharge of gases with the resultant crater becoming choked with rocks collapsing from the steep side walls. In a repeat of the story at the Lion's Head, basaltic magma intruded up through the broken rubble and crystallised in this case to yield the basalt capping Crow Hill. Once again the post-explosive stage probably saw ponding of the magma within the crater as a lava lake. However, in both vents the many tens of metres of rock that originally overlay the present summits have been lost by erosion. Several lavas are present at the far south-west of the vent with well-bedded sedimentary rocks intercalated between them. These sediments suggest the periodic formation of a crater lake.

Three substantial basaltic masses within the Lion's Haunch Vent are regarded as intrusive. Of these, the Samson's Ribs mass, along the south-western contact of the vent, shows spectacular columnar jointing. From the top of the 30 m high cliff above the road west of Duddingston Loch, the SSW dipping columns steepen as they are followed downwards from around 60 to 75° before turning out at much shallower angles to lie almost perpendicular to the rock face. The lower columns appear to have grown in response to cooling against an almost vertical side wall, whereas the upper portions grew in response to heat loss from a sub-horizontal upper surface, possibly another lava lake within the Lion's Haunch crater. The second basalt mass surrounded by vent breccia forms Crow Hill, the summit area of the Lion's Haunch. Dunsapie Hill, in the north-eastern extremity of the vent, represents a third principal intrusion. This is roughly cylindrical in form and, like the Samson's Ribs basalt, is thought to have been intruded along the contact zone of the vent, between the volcanic breccia and the Whinny Hill lavas. According to Black's interpretation, lavas 5, 6, 7 and perhaps those above, were all erupted from the Lion's Haunch Vent. Since, as we have seen, the land surface around Edinburgh at about the time lavas 1 and 2 were erupted was still little above sea-level, the volcano crowned consecutively by the Lion's Head and Lion's Haunch summit craters could not, at its maximum development, have risen a great deal higher above sea-level than the present-day summits.

Not only were the strata in the Dry Dam succession water-lain but the ash horizons underlying lavas 3 and 4 show indications of having accumulated in shallow water. In view of this the whole volcanic edifice was almost certainly very low lying and close to sea-level, although cones above e.g. Castle Rock, Pulpit Rock, Lion's Head and Lion's Haunch Vents would have stood substantially higher. The suggestion that, at its maximum, the volcano may have risen to about 1,000 m above sea-level, with a cone-base of up to 5 km in diameter, could well be an exaggeration.

To the south of the two culminating vents the Whinny Hill lava and ash succession reappears on the north side of Duddingston Loch. In these southerly exposures lava 1 and the ashy sedimentary strata of the Dry Dam can be readily correlated with those of Whinny Hill. Above these is a thick, coarse-grained fragmental (pyroclastic) unit, poorly exposed in the vicinity of Duddingston village, within which two thin lavas can be discerned. This unit is overlain by several lavas that possibly match lavas 8 to 13 of Whinny Hill. However, there is a strong asymmetry between the volcanic successions seen around Whinny Hill and Duddingston: fragmental, ashy deposits are much thicker in the latter, appearing to take the place of lavas in Whinny Hill. This asymmetry has been ascribed by George Black to strong northerly winds that caused greater accumulation of air-fall on the more southerly flanks.

A basalt plug, on which the old Duddingston church stands, lies outside the Lion's Haunch Vent. It is ovoid in plan (*c.*250 m across) but the only significant outcrop is that on the shore of Duddingston Loch. As with other basaltic plugs in Edinburgh (Castle Rock, Pulpit Rock and perhaps Dunsapie Hill) the Duddingston plug could well represent a blocked conduit that supplied surface eruptions.

In many popular writings of the geology of Edinburgh the term 'Edinburgh's Volcano' appears, giving the impression that there was only one. On the contrary we see that there was a complex volcanic field, a miniature reflection of the Clyde Plateau to the west, developed in the Edinburgh district from the products of a cluster of fairly small basaltic volcanoes. Since the whole lava field disappears from view eastwards beneath the later sedimentary strata of the Midlothian Coalfield, there may well have been other participatory volcanoes of which we are currently ignorant.

Within the western confines of Holyrood Park the eastward-dipping sedimentary rocks underlying the lavas and pyroclastic rocks consist of

shallow-water sandstones and limey mudstones. These strata contain two sills, the Heriot Mount–St Leonard's Sill and the Dasses Sill. Although both of these intrusions are thought to have been contemporaneous with the extrusive activity, the much more prominent Salisbury Crags Sill, which lies approximately mid-way between these two, was injected much later (*Fig. 8.8*). The sill is of international importance in the history of the science. In the late 18th century James Hutton, a famous doyen of the Scottish Enlightenment, realised that hot molten rock (magma) had been injected into sedimentary strata. Tradition maintains that Hutton used the site at the base of the sill near its southern end ('Hutton's Locality') to demonstrate its intrusive nature and to refute the widespread competing belief that whinstone (i.e. dolerite) and basalt were marine precipitates. The Salisbury Crags Sill attains a maximum thickness of some 40 m but thins southwards towards its contact with the Lion's Haunch Vent. The precise age of the sill is unknown. Geological relationships show it to be younger than the Lion's Haunch Vent while its composition allies it more closely to the later Carboniferous magmatism discussed in Chapter 7. The sill is cut by a small

Fig. 8.8 Salisbury Crags sill with Whinny Hill (back-ground left) and Arthur's Seat (centre background).

E–W dyke belonging to the $c.300$ Myr swarm discussed in the last chapter. Consequently, on the basis of these observations and its chemical composition, it probably dates from between 330 and 315 Myr. Although the presence of prominent gas vesicles near its upper surface shows it to have been intruded at shallow depths (one or two hundred metres?) there is no reason to suppose that it had a surface connection.

The intrusion of sills at shallow depths is likely to have been instrumental in the episodic inflation of near-surface sedimentary strata, producing emergence and allowing sub-aerial weathering and plant growth. Since plant fragments are commonly encountered in the ashy layers we may envisage the volcanic hills as having been forested during long periods between the occasional eruptions. Subsidence and inundation following lava 1 allowed deposition of the well-bedded ashes with intercalated lagoonal sediments seen in the lower part of the Dry Dam to the north and the lower part of the 'lower ash' at Duddingston. The plant fragments within the Dry Dam sequence were probably washed down into shallow waters from the adjacent forested volcano flanks. The (poorly exposed) Crags Vent that lies west of the Lion's Head, which was probably surmounted by a basaltic cinder cone approaching 1 km in diameter, may have developed fairly early, possibly contributing to the ashes of the Dry Dam. The whole of the Whinny Hill, Lion's Head–Lion's Haunch area is clearly very shallowly dissected and as mentioned above, the larger basaltic outcrops within these two vents may have had surface expressions as confined lava lakes.

If the Castle Rock volcano fed lava 1, it was erupting at an early stage, certainly before the Lion's Head and Lion's Haunch Vents. In contrast the Duddingston plug could represent the conduit for a parasitic volcano developed at a late stage on the south-eastern flanks of the main edifice. The Edinburgh volcanoes may thus have been distributed along a WNW–ESE lineament, $c.2.5$ km long, exhibiting a very generalised migration of activity over time from Castle Rock in the WNW to the Duddingston plug in the ESE. Although still debatable, it is thought that the lavas composing Craiglockhart Hill in the south of Edinburgh could be still older than the Castle Rock

East Lothian

While volcanism was occurring in the Edinburgh district, larger eruptions were taking place at much the same time to the north-east in East Lothian, in the general area of North Berwick and Haddington. The volcanic rocks of East Lothian outcrop between the extrapolation of the main Southern

Upland Fault and the associated, sub-parallel Dunbar–Gifford Fault (*Fig. 2.1*). This represents a down-faulted area of Carboniferous deposition which strictly lies within the Southern Uplands terrane rather than within the Midland Valley proper. Nonetheless the siting of the volcanoes, whose eroded relics form much of the scenic coastline (*Fig. 8.9*), was almost certainly dictated by faulting related to the southern boundary of the Midland Valley. The lava pile built up in this region (over 500 m thick) dips westwards towards the Midlothian Coalfield syncline. Since, as we have seen in the previous section, the Edinburgh volcanic strata dip eastwards into this syncline it is possible that they and the East Lothian lavas were originally all part of one volcanic field. On this hypothesis it could be that we do not see the intervening part because it now lies deeply buried beneath a stack of younger Carboniferous strata occupying the axis of the downfold. In other words if mining had continued deep below the coal-bearing strata it would have encountered early Carboniferous lavas. Frankly, we do not know although the general opinion is that the contemporary East Lothian and Midlothian volcanic fields were quite separate and not contiguous. Since much of East Lothian is agricultural and the rocks are poorly exposed, we can only guess where most of the volcanoes themselves were sited.

The first products were basaltic ash layers from explosive eruptions, just as for the Clyde Plateau and the Arthur's Seat–Calton Hill sequences, generated when the first arriving magmas encountered surface water. Scarce, thin sedimentary strata interlayered with these ashes are non-marine in character and the waters appear to have been lagoonal rather than part of the open sea. North Berwick is built on these pyroclastic deposits and the rocks are superbly exposed at low tide along the foreshore for over 6 km from west of North Berwick east to beyond Tantallon Castle (*Fig. 8.10*). As we see them now the ash fragments are either dull greenish or rusty red-brown, typically a few millimetres across and usually separated by white calcite that acts as a cement. When pristine and newly formed most of the basalt pieces would have been of shiny black basaltic glass, filled with tiny gas bubbles. As has been explained, glasses are delicate ephemeral materials that normally have short lifetimes in geological terms. They dissolve in and react with water, losing their glassy character and becoming crystalline on a fine scale, with growth of minerals of the clay and chlorite families. Chlorites are typically bluish-green and it is the chlorites that colour much of the outcrops. In more oxidising environments, at least some of which is believed to have been caused by penecontemporaneous weathering, the red iron oxide, haematite, was generated, giving the rusty coloured outcrops.

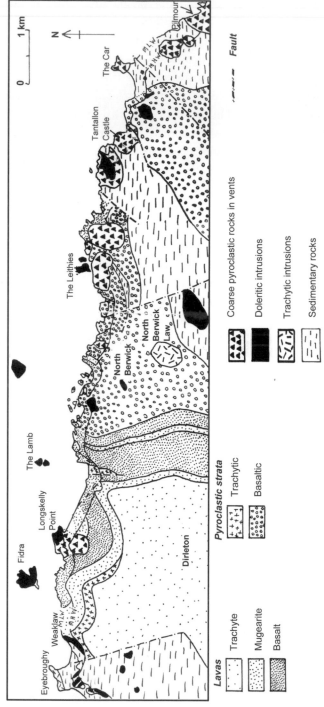

Fig. 8.9 Geological map of part of the East Lothian coast including North Berwick. *(After McAdam and Clarkson, 1986)*

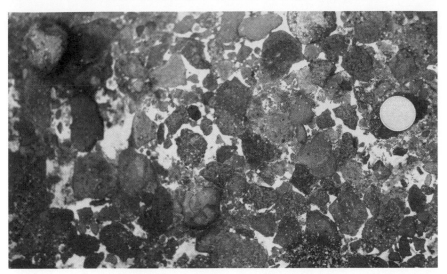

Fig. 8.10 Basaltic ash deposits on the foreshore, North Berwick. The white interstitial material is secondary calcite cement. Coin 25 mm diameter.

Calcium carbonate, precipitated as the mineral calcite from the (probably hydrothermal) waters that later percolated through the uncompacted ashes, led to their cementation (*Fig. 8.10*).

As with the cases discussed above, intrusion of basalt magma at shallow crustal levels was probably responsible for uplift and retreat of the waters and consequently eruptions spilled their lavas across emergent land. The early lavas were generally basaltic although some, rather more evolved (i.e. cooler and with lower magnesium and higher sodium and potassium contents) were compositionally intermediate between true basalt and trachyte (as discussed in Chapter 3) and are known as mugearites (after Mugeary, a small hamlet on Skye). Four of these lavas, which dip west at *c*.20°, are very well exposed between North Berwick harbour and the paddling pool some 100 m to the east. A thin (*c*.4 m) and intensively altered initial lava flow was overlain by several metres of ash beds before emergence was finally complete. The first of three lavas that subsequently erupted was a basalt flow (*Fig. 8.11*) while the second was a mugearite, with a distinctive platy fracture on which there is a well preserved upper surface showing the original sharp and blocky top. The vesicles (now amygdales) in this flow top have been stretched out into elongate tubular forms and individual lava blocks have been rotated as they were carried along by the flow. The original surface clearly experienced such minimal weathering before it was over-ridden by the third (basaltic) lava flow that this final eruption must have followed hard on the heels of the previous one.

Fig. 8.11 The lowest of a group of three lavas at North Berwick Harbour. The prominent red horizon below the basalt lava is of lateritised ashes, severely weathered and oxidised in a tropical climate.

The youngest lavas in East Lothian were distinctly different in being trachytes. Whereas their predecessors had been moderately fluid lavas little over 10 m thick, the trachytes appear to have been much more viscous, attaining thicknesses of 20 m or more. These form the Garleton Hills where they give rise to some fine north-facing escarpments. They attain their greatest thickness in the Haddington area and, because of the general stickiness of such lavas, the probability is that their eruptive focus lay in this district although, as yet, it has not been possible to identify it.

On the coast of the Firth of Forth, west of North Berwick, a mugearite lava is exposed. This may or not be the same as that seen at North Berwick itself but, again it is remarkable in having a ragged, jagged *aa*-type flow top preserved in a state little modified from that when it finished flowing about 340 million years ago. It is overlain by a succession of finely bedded trachytic ashes. The inference must be that the mugearite lava had scarcely been erupted before a devastating explosive event occurred that showered the landscape with angular debris, thus leaving us with one of the most untouched and pristine lava surfaces in Scotland. The ashes are notably coarse, with angular fragments up to several centimetres across, suggesting that the eruptive site lay at no great distance. The ashes are themselves overlain by the thick trachyte lavas described above.

Somewhere, probably deep in or at the base of the crust, maturation of large basaltic magma bodies by the process of fractional crystallisation gave rise to a substantial volume of trachytic magma. In the final stages of the volcanism this low-density (buoyant) magma ascended to a near-surface level, where the overlying rocks were no longer strong enough to cap the high-pressure water vapour and other gases previously held in solution. With the fracturing of the roof the gases were explosively released, possibly in a plinian eruption column (*Fig. 4.4*). The underlying trachyte magma body, having now lost most of its gas content, then erupted as thick and stodgy lavas. It would be expected that, accompanying this violent, cathartic evacuation of gas, magma and wall-rock fragments, a caldera collapse resulted. If so, such a structure has not yet been detected beneath the farmlands of East Lothian although it may be that the trachyte lavas of the Haddington area were ponded within such a caldera depression.

I have not written of the principal landscape features of East Lothian, namely the pyramidal hill of North Berwick Law nor the upturned pudding-basin topography of Traprain Law (*Fig. 8.12*). Of the several islands close offshore in the Firth of Forth (mostly due to basaltic sill-like intrusions) the most famous is Bass Rock, with its history of hermitage, penal colony and currently bird sanctuary. All three are composed of trachyte intrusions. Both North Berwick Law and Bass Rock have the form of roughly cylindrical necks, about half a kilometre in diameter. Both are regarded as sub-volcanic structures that fed surface eruptions. At Traprain there are three trachytic intrusions, the largest of which composes the Law. This has up-arched the country-rock sandstone and its overlying volcanic strata and rocks on top of the hill (the ancient stronghold of the Votadini tribe) are vesicular, signifying loss of gas at low pressure close to the surface. The Traprain trachytes are accompanied by two small steep-sided basaltic necks and it is not improbable that the whole ensemble represents a (fairly shallowly) eroded volcanic complex that involved both trachytic and basaltic magmas.

The Bass Rock, North Berwick Law and Traprain trachytes all differ in composition from the trachytic lavas of the Haddington area in being less

Fig. 8.12 View of Traprain Law, East Lothian.

siliceous but somewhat richer in alkali metals (potassium and sodium). Accordingly, while we cannot identify the spot from which the lavas were erupted, we can also find no lavas that can be matched to the intrusions that probably did underlie volcanoes! The answer to the latter problem is that the intrusions are likely to have fed very localised bulbous domes of lava that never spread far away from their feeder and were destroyed by erosion in a comparatively short space of time. Close modern analogues, are to be found in central France in the Chaîne des Puys of the Auvergne (*Fig. 8.13*). The question why the younger East Lothian trachytes are deficient in silica relative to the observed lavas is unanswered. It could be that the earlier trachyte magmas reacted with and assimilated more of the silica-rich country-rocks through which they passed and that, for some reason, the later magma batches were less affected. The answer may well be supplied by future research.

Silica-poor trachyte intrusions very much akin to those of East Lothian occur elsewhere in the Midland Valley and are also regarded as having similar early Carboniferous ages. The most striking of these from a scenic viewpoint is Loudoun Hill in Ayrshire, east of Kilmarnock to the north of the A71 (*Fig. 8.14*). Loudoun, a battlefield site in the wars of Scottish independence, forms a resistant upstanding feature reminiscent of North Berwick Law, to

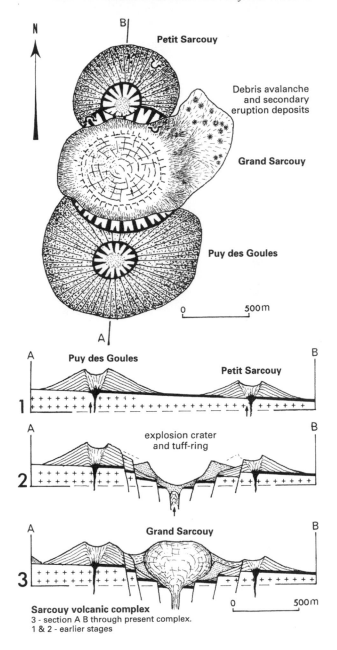

Fig. 8.13 Diagrams illustrating the evolution of one of the volcanoes in the Chaîne des Puys, Massif Centrale, France. Early basaltic activity was followed by the extrusion of a viscous trachyte dome, Le Grand Sarcouy. Some of the early Carboniferous trachytes in Scotland may have originated in this manner. *(GOER de HERVE A. de, BOIVIN P., CAMUS G., GOURGAUD A., KIEFFER G., MERGOIL J., VINCENT P.M. – 1994 – Volcanology of the Chaîne des Puys, 3e éd (English version). – Parc Naturel Régional des Volcans d'Auvergne édt, 127.p.)*

Fig. 8.14 Loudoun Hill, Ayrshire. An early Carboniferous trachytic plug. *(C.H. Emeleus)*

which it is compositionally close. Again, the chances are that the trachyte magma supplied through the Loudoun neck fed a viscous extrusive dome that grew rather like a blob of toothpaste extruded from its tube.

The rocks of the East Lothian coast provide an insight into an ancient equatorial world of sluggish rivers, lakes and lagoons. This landscape was subject to short-lived, violent eruptions from small volcanoes that stood above an otherwise flat landscape. The wide variety of early land plants which flourished on the flanks of these volcanoes makes the region one of international palaeobotanical importance. Lenses of magnesian limestone, interlayered with the ash beds, are thought to have been deposited from solution as near-shore lagoons evaporated during periods of aridity. Possible algal growths have been discerned in some of these limestones. Thin sedimentary layers of limey muds (marls) and sandstones are also interbedded with the ashes. While these contain no fossils of marine organisms, the remains of small crustaceans (ostracods) and fragments of plant fossils are fairly common. All the evidence points to a non-marine environment, although it was clearly one with abundant shallow water not very far from the sea.

Some fourteen volcanic vents have been recognised along the East Lothian coast (*Fig. 8.9*) and it may be presumed that many more lie both

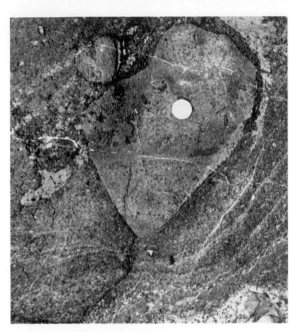

Fig.8.15 Ashy strata within the 'Pilmour Volcano', East Lothian coast. A large block was a projectile that has impacted into unconsolidated ashes. The reddish colours are all due to much later oxidation effects. Coin 25 mm diameter.

out to sea and inland beneath drift deposits. It is often difficult to make a distinction between the mapped vents and the surrounding bedded ashes and sedimentary rocks composed of both volcanic and non-volcanic particles. However, the latter represent widely distributed fragmental material on and around the volcanic cones which were subject to water sorting and admixture with fluviatile detritus, whereas the vent material is the coarser and more chaotic debris that collected as fallout and talus on the steeper inward-facing slopes of the cones. Faulting and slumping of waterlogged debris added to the complexity. The vents along the East Lothian coast appear to represent shallowly dissected tuff-rings, the diameters of which rarely exceeded 1 km and were often significantly less. Their appearance would have been similar to that shown in *Fig. 7.5*. Eruptions would have been short-lived and brutal, yielding pyroclastic products composed mainly of broken fragments of near-surface rock formations together with blobs of new lava, including at times lava bombs of substantial size (*Fig. 8.15*). Sedimentary material within some of the vents suggests the probability that lakes ('maars') formed within the craters. Plant fossils suggest that the emergent volcano flanks provided habitats for ferns, horsetails (equisitales) and clubmosses, fragments of which are common in the ashes and which are well preserved in places, as at Oxroad Bay below Tantallon Castle east of North Berwick and at Weaklaw some 5 km west of North Berwick (*Fig. 8.9*).

Fig. 8.16 An angular fragment of mantle peridotite in the pyroclastic rocks of the Weaklaw Vent, East Lothian. The whitish material with flow layering that surrounds the peridotite was originally composed of basaltic glass. This glass was hydrothermally replaced by microcrystalline carbonates and clays that confer the pale colour. Coin 25 mm diameter.

The Weaklaw Vent is, in some respects, reminiscent of the Heads of Ayr Vent described above, on the Clyde coast. As at the Heads of Ayr, we are probably seeing relics of the surface cone that has fallen back, through circumferential (or ring) faults developed when explosive steam eruptions had come to a close. As at Heads of Ayr, rapidly rising silica-poor basaltic magmas have wrenched off pieces of their mantle (and lower crustal) walls and flushed them up from depths of over 30 km to surface levels. Chunks of mantle peridotite, up to grapefruit-sized, arrived at the top wrapped in coatings of highly vesicular, glassy basalt. However, the glasses and virtually all of the original peridotite minerals have been replaced by carbonate, chlorite and clay minerals from reactions with carbonate-rich aqueous fluids exhaled through the fragmental vent filling during an extended history of cooling (*Fig. 8.16*).

The ruined Dunbar Castle was built on top of one of these East Lothian rubble-filled vents. About halfway between Dunbar and North Berwick is the headland beside the estuary of the East Lothian Tyne bearing the name St Baldred's Cradle, after one of the local hermits. St Baldred's Cradle is also the site of an old volcano. Although not precisely dated it was probably of early Carboniferous vintage. Vestigial remains of a down-faulted, well-bedded basaltic ash-cone are here cut by a basalt plug with steeply inward-dipping contacts. It is tempting to suppose that this outward-flaring basalt mass opened upwards into a lava lake that was originally ponded within the confines of a tuff-ring. In any case, this well-jointed basalt plug affords superb views of Bass Rock and other volcanic features on either side of the Forth estuary.

The Southern Uplands and the Scottish Borders

If we go further into the Border region we again find evidence that volcanism took place at about the same time as the much more productive activity in the Midland Valley that yielded the Clyde Plateau and other lavas. The beautiful Eildon Hills that rise above the River Tweed near Melrose owe their origin to a suite of trachytic and rhyolitic sills intruded into the uppermost Devonian sedimentary strata. While no surficial products survive a small basaltic hill (Mid-Hill) in the Eildons may have fed an overlying volcano. Rising immediately south of Melrose is a green hill, adjacent to the Borders General Hospital, underlain by a large vent (the Chiefswood Vent). This is occupied by an unbedded and unsorted mass of a finely broken mixture of sedimentary rocks derived from the underlying and surrounding Silurian slates and Devonian sandstones, with igneous fragments of trachyte and rhyolite. The vent originated through the interaction of trachytic and/or rhyolitic magma with surface or near-surface waters. Consequent violent steam explosions blasted both the country-rocks and near-surface trachytic/rhyolitic intrusions to small pieces. A broad crater (3 x 1 km) may be envisaged surrounded by a tuff-ring that was probably occupied by a crater lake. The conduit became blocked with fall-back talus composing the vent fill. That rhyolitic magma was in close proximity and remained available for intrusion is evidenced by a rhyolite ('quartz porphyry') dyke that transects the vent. The mottled tawny-brown to greenish-grey vent rock makes an attractive building stone, as was recognised by the 12th century Cistercian monks who used it extensively in the construction of Melrose Abbey.

Going back now another fifteen to twenty million years we close in on the start of the Carboniferous Period at *c.*359 million years ago, to a time before Scotland moved north across the equator. The tropical rain forest conditions that characterised central Scotland in the mid- to late-Carboniferous were still tens of millions of years into the future and the climate, although still equatorial, was distinctly drier. South and central Scotland had a subdued landscape. The Midland Valley and northern

Fig. 8.18 Smailholm Tower, near Kelso, standing on outcrops of lavas and intrusions.

England were lowland regions of meandering, braided rivers with sandy and gravelly flood plains separated by a belt of higher ground, the putative Southern Uplands. We may imagine the plains to have supported a modest flora of ferns and other humble plants. The vertebrate fauna would have included fish and salamander- or newt-like amphibians. Just as the Scottish Carboniferous strata are famed from the occurrence at East Kirkton of primitive vertebrates that were just making the revolutionary step from being water-bound to a more fully terrestrial lifestyle, so Scotland currently also holds the honour of hosting the earliest known vertebrate with four legs adapted for crawling on land. This fossil, which dates from around 350 Myr, was found near Dumbarton in shallow-water sediments deposited prior to the Clyde Plateau volcanism. It was a thick-set, crocodile-like creature (*Pederpes finneyae*), which represents a 'missing link' between the more advanced fish of the preceding Devonian Period and the amphibians that became important life forms in the Carboniferous swamps.

Although the passage from the Devonian to the Carboniferous Periods in northern Britain was undramatic and marked by no obvious change, volcanoes began to appear early in the Carboniferous at about 350 Myr, perhaps several million years in advance of the far more serious and extensive activity considered above for the Clyde Plateau, Edinburgh region and East Lothian. The volcanism took place along a narrow zone paralleling and approximating to the English–Scottish border. This zone, which extends from the northern margins of the Solway Firth east-north-east towards Kelso and Coldstream, was characterised by down-faulting and down-folding towards the south, thus presenting a southward-facing escarpment towards a developing fluviatile basin of the Northumberland Trough. A more or less continuous line of small basaltic volcanoes would have marked this broken hinge-zone. Lavas from these, never accumulating to any great thickness, are now seen as a discontinuous strip of outcrops from south of Criffel (near Caulkerbush) to between Echelfechan and Locherbie. Travellers may see a small flat-topped hill, topped by an iron-age fort (and subsequent Roman camp), at Birrenswark (or Burnswark) on the north-east side of the M74, which is underlain by these lavas. The outcrops persist north-eastwards beyond Langholm and the thickest and most extensive crop of lavas belonging to this preliminary activity occurs in the Kelso–Coldstream area. Here the lava sequence was subsequently bent into a downfold and the outcrop now has the shape of a tilted 'U' with the open end towards the north-east. Noteworthy among these Kelso lava outcrops are those at Stichill and Hume, a few kilometres north of Kelso.

Fig. 8.19 Basaltic cones of recent Cainozoic age in the Antsirabe region, Madagascar. The early Carboniferous landscape of the Kelso, Melrose and Langholm district probably had a similar appearance. *(W.J. Wadsworth)*

While it is difficult (or impossible) to assign any of these lavas to a particular volcano, there are several dozen volcanic necks in the region of Langholm, Hawick, Jedburgh and Melrose, thought to be the eroded conduits of these earliest Carboniferous volcanoes. Many form notable hills and a particularly handsome one is Ruberslaw, to the east of Hawick. Although the eroded profile of this hill has a volcano-like form, the viewer must remember that the actual volcanic cone that this conduit supplied lay some hundreds of metres above the present summit and, like all the others, was long ago wiped clean by the forces of erosion. Whereas Ruberslaw is a basalt-filled plug, the Minto Hills, 5 km north of Ruberslaw, appear to be entirely filled with fragmental (agglomeratic) rocks. Smailholm Tower, a picturesque 15th century fortification WNW of Kelso, is built on a basaltic plug cutting the Kelso lavas (*Fig. 8.18*). The upland landscape in the Borders during these early Carboniferous times may have resembled the late Cenozoic volcanic scenery of Madagascar shown in *Fig. 8.19*.

Chapter 9

Volcanoes in the Old Red Sandstone Continent

Turning the pages of history to beyond the Carboniferous we come to the Devonian Period, which derives its name from the fact that strata of this age were first recognised in the south-west of England. The timing of the start of the Devonian remains debatable. According to different opinions the onset may have been as early as 416 million years ago or as recent as 409 million years ago, while there is rather more certainty that it ended at around 359 million years ago. As outlined in the previous chapter, the strata in Scotland that were laid down at the close of the Devonian pass up, without interruption, into those of the Carboniferous. Globally, however, there was a very significant change in the fossil fauna that marks the passage from Devonian to Carboniferous. Just as the uppermost Devonian strata grade upwards into the lowest Carboniferous strata, so at the base of the Devonian in North Britain there is nothing to denote the changeover from the rocks of the preceding Silurian Period.

The Old Red Sandstone Continent and its formation

Much of the Devonian and Silurian strata in SW England and Wales, where the periods were first defined in the 19th century, were deposited on sea floors. These rocks are consequently well endowed with marine fossils and it is the evolutionary changes in these that were critical in distinguishing one period from another. In contrast, the late Silurian and early Devonian rocks in Scotland are non-marine and lack these vital indicators. Much of northern Europe, together with Greenland and North America, were emergent at the time and were constituent parts of a continent, now formally referred to as Laurussia but which has for long been referred to as 'the Old Red Sandstone Continent'. Because of the problem of defining the Silurian–Devonian boundary, it has been convenient to lump the late Silurian strata together with those of the Devonian under the single

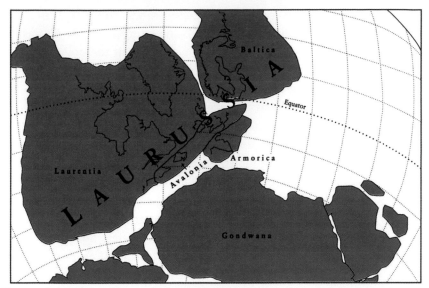

Fig. 9.1 Palaeogeographic reconstruction of Laurussia in late Devonian time (*c.*380 Myr). *(After Geological Conservation Review, 2002, eds. D. Stephenson et al.)*

heading of 'Old Red Sandstone'. Sandstones are certainly abundant, although commonly they grade into much coarser-grained conglomerates composed of cobbles or boulders. The rocks are typically reddish from the presence of the red iron oxide haematite, deposited between the particles by percolating waters in an oxidising environment. The word 'old' is inserted to distinguish these Siluro-Devonian rocks from the somewhat similar 'New Red Sandstone' rocks laid down very much later during the Permian and Triassic Periods. In the latest Silurian and earliest Devonian times all of Scotland (or all those components that were subsequently to become Scotland) lay towards the south-east margin of the Laurussian continent, south of the equator (*Fig. 9.1*).

It would appear that for the past 416 million years all the constituent parts (or terranes) of Scotland were joined together, to form a single continental unit. But, as we go further back into the Silurian, matters were more complicated and it is necessary to review a little of the earlier history before describing the volcanism. It is now widely accepted that in the Lower Palaeozoic an ocean existed to which the name 'Iapetus' has been given and, as outlined in Chapter 2, the Iapetus Ocean first came into existence some time around 600 million years ago. During its early history the ocean grew and then, when consumption along subduction zones exceeded the rate at which new oceanic lithosphere was being created, it began to contract and eventually disappeared. Although simplistically one may say that Iapetus opened and closed like a concertina, the closure was complex. It involved a prolonged oblique collision of the opposing continental margins as the ocean floor was swallowed by subduction. Closure of the ocean involved a sequence of mountain-building (i.e. orogenic) events that took place throughout much of the Ordovician and Silurian periods.

The name Caledonian was first applied to the resultant mountain belt by Suess in 1885 from his studies of the pre-Old Red Sandstone rocks in Scotland. He and his contemporaries realised that the highly folded rocks beneath the Old Red Sandstone strata represent the eroded remains of a once great mountain range that had lain across Scotland, Ireland, Wales, northern England and Norway. Through the subsequent painstaking work by generations of geologists, a picture of the events leading up to the formation of these mountains has come into increasingly sharp focus. The realisation that these events involved the closure of an ocean constituted a major intellectual breakthrough without which the whole process would have remained mysterious. The Iapetus Ocean had been created through the rifting and break-up of an earlier great continent, Rodinia. The disrupted pieces of Rodinia separated as the new ocean expanded and it was around 510 Myr, in the later part of the Cambrian Period, that the ocean is thought to have attained its maximum width, perhaps as much as 5,000 km. As a result of its subsequent contraction, three continental fragments, called Laurentia, Avalonia and Baltica, were brought into collision (*Fig. 9.1*). Consumption of the ocean took place along a complex pattern of subduction zones and Laurentia, Baltica and Avalonia were drawn together on a mutual collision course. Laurentia incorporated what became North America, Greenland and NW Scotland. Baltica included much of what is now north-western Europe (including Poland, NW Russia, the Baltic States and Scandinavia) while Avalonia included some of the more southerly regions of western Europe, England, Wales, southern Ireland and parts of Newfoundland. While these three appear to have been the principal components of the new continent, smaller volcanic arcs were probably also involved in its generation. *Fig. 9.2* shows a reconstruction of the palaeogeography at around 430 Myr, showing 'Scotland' lying in tropical latitudes some way south of the Equator.

After the demise of the Iapetus Ocean, that part of the resulting new continent destined to become Scotland would have been spectacularly mountainous. Most of the pre-Old Red Sandstone rocks of the Grampian and Northern Highlands acquired their deformed and metamorphosed character within this mountain range in which the major uplift took place between 460 and 430 Myr. We do not know how high these moun-

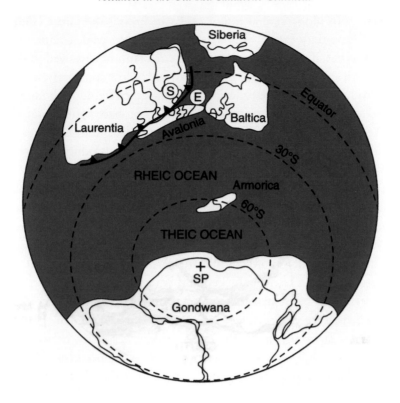

Fig. 9.2 Possible configuration of Laurentia, Avalonia and Baltica in the middle of the Silurian Period. 'S' marks the location of 'Scotland' while 'E' marks the location of 'England' and 'SP' indicates the South Pole. While a single north-westerly directed subduction is indicated (heavy line with arrows) the plate tectonic boundaries are likely to have been more complicated. *(After G.J.H. Oliver, in* The Geology of Scotland, *4th edition, N.H. Trewin, ed.)*

tains were but the inferences are that they were of Alpine and possibly even Himalayan proportions. As a generalisation one may consider that throughout its subsequent history the great Caledonian mountains were being remorselessly worn down by the action of rain and rivers. Erosion would have been most rapid when the mountains were at their highest, slowing down progressively as they were reduced to pale shadows of their former glory by the time the Devonian Period ended. According to one estimate some 20 km thickness of overburden had already been stripped by 430 Myr. With the passage of time the rivers ran slower, the flood-plains and meanders became broader and the transported sediments changed from coarser to finer grades.

The orogeny had no sooner peaked than a long period of relaxation, faulting and extension affected the mountain belt. This was accompanied by lateral shifts along several major faults including the Great Glen and Highland Boundary Faults. We have already encountered (Chapter 6) the convention that, if the rocks on the far side of a fault have moved to the left in relation to an observer, it is a 'left-lateral' fault while conversely, if motion appears to have been to the right it is a 'right-lateral' fault. Although most of these fault movements were left-lateral, some had right-lateral shifts. Complex rotations of faulted blocks also took place and, accompanying these adjustments, magmas were generated at depth producing intrusions on a wide variety of scales and gave rise to volcanoes.

By the end of the Devonian the Midland Valley had been structurally defined by the Highland Boundary Fault on the one side and the Southern Upland Fault on the other. However, studies of the sedimentary rocks (mainly river-borne sandstones and conglomerates) in the Midland Valley indicate a much more complicated picture. Such studies tell us the directions in which the rivers flowed and where the sediments came from and, in so doing, present some serious problems. For example, whereas huge quantities of rock were eroded from the Grampian Highlands (part of the Caledonian belt), the debris does not start to be seen in the Midland Valley until very late in the Devonian. It has been concluded that there *must* have been a basin to the south-east of the Grampians, which received this material in Silurian and earlier Devonian times but which is no longer to be seen. The missing tract is deduced to have been over-ridden by the Grampians, thrust south and east over a low-angled fault and consequently now hidden from view. Similar problems arise in the south of the Midland Valley, where again a depositional basin that received detritus from the eroding Southern Uplands is no longer visible. So, both the Highland Boundary Fault and the Southern Upland Fault may have had earlier histories as low-angled thrust faults, encroaching on either side of what is now the Midland Valley. As a further complication, large lateral motions, of perhaps tens of kilometres, may also have been involved as well as over-riding thrusts. Some of the volcanic pebbles in the Midland Valley conglomerates constitute our only tangible evidence for volcanoes that were once present in these marginal belts.

Despite the uncertainties in the age dating, as well as in the estimate of latitudes from studies of rock magnetism, there is general agreement that the climate on the Old Red Sandstone Continent was warm to hot, semi-arid and with some thoroughly arid periods. The Caledonian mountains rose above lowlands which supported braided river systems, wide flood plains and, from time to time, lakes. Volcanoes were widespread and the principal lava successions are now preserved in the Southern Uplands, the Midland Valley and the Highlands south of the Great Glen Fault. Minor occurrences are also present in the Shetlands and Orkneys. *Fig. 9.3* shows the main outcrop distribution and also a palaeogeographic reconstruction showing the rough pattern of highlands, lowlands and volcanoes. Some of the late Silurian to Devonian intrusions may be suspected to have underlain volcanoes, as pebbles in conglomerates have come from volcanic sources that are no longer accessible, either because they have been covered by over-riding thrust faults, as indicated above, or because they now lie offshore.

Life around the Old Red Sandstone volcanoes

An invaluable insight into the sort of plants and animals that had begun to colonise the newly formed Old Red Sandstone (or Laurussian) Continent has been gleaned from some remarkable fossils in Aberdeenshire. These occur in silica-rich rocks (cherts) near the village of Rhynie which are deduced, from palaeomagnetic data, to have been formed in a sub-tropical environment around 28° S. The cherts formed as hot-spring deposits in a narrow NE–SW-trending basin, within which river and lake sediments accumulated, together with some andesitic ashes and lavas (*Fig. 9.4*). The waters from which the cherts precipitated had been heated by the rising andesite magmas. The cherts have acquired global renown because of the perfect preservation of the terrestrial and freshwater communities of plants and animals trapped in them. It appears that chert precipitation (i.e. silici-fication) took place immediately after the death of the organisms, thereby preventing decay. The plants include moss-like vascular species (i.e. with water-conducting cells) as well as more primitive species of algae, fungi and lichens. Small creatures living amongst this flora were also entombed and silicified. These include myriapods (ancestral forms of millipedes and centipedes), small arachnids (mites) and primitive crustaceans. The Rhynie cherts have, in consequence, great scientific value for their evidence of some of the world's earliest terrestrial plants and animals. They provide a metaphorical shaft of sunlight illuminating an ancient ecosystem compa-rable to that shed on the mid-Carboniferous by the East Kirkton deposits described in the previous chapter. Humble plants, as exemplified by those at Rhynie, first started to colonise the land in late Silurian times and evolved vigorously throughout the Devonian Period. Consequently we may think of the watered lowlands as having been verdant for the first time in the history of the planet.

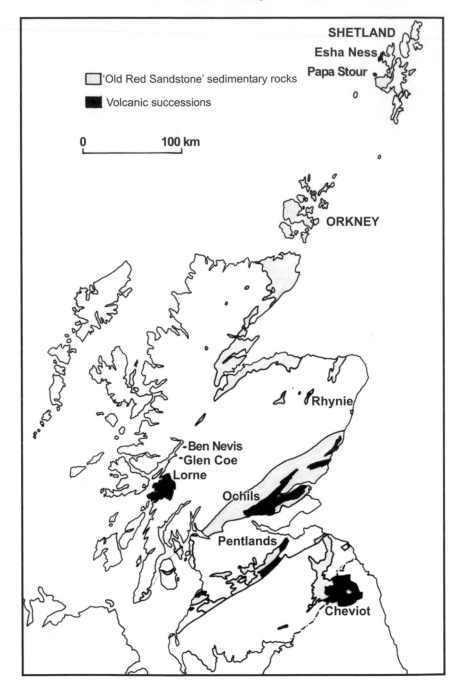

Fig. 9.3a Present distribution of late Silurian–early Devonian volcanic rocks in Scotland. *(After N. H. Trewin and M. E. Thirlwall, in The Geology of Scotland, 4th edition, 2002)*

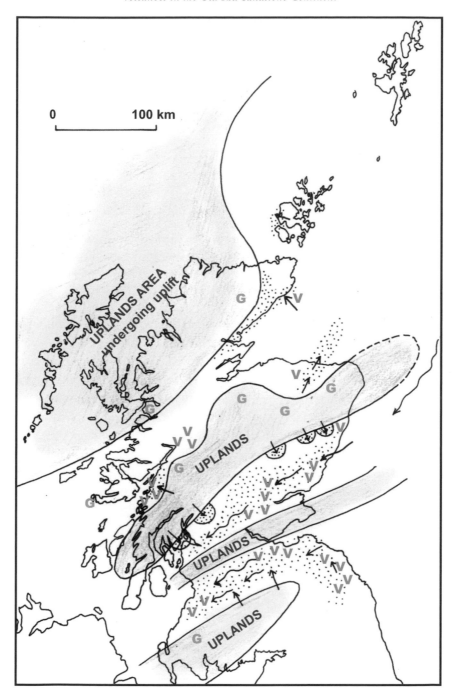

Fig. 9.3b Inferred palaeogeography of Scotland in late Silurian–early Devonian times. Arrows show direction of rivers and sediment transport. V indicates the areasa of preserved volcanic rocks. G shows area of emplacement of granite. *(After N. H. Trewin and M. E. Thirlwall, in The Geology of Scotland, 4th edition, 2002)*

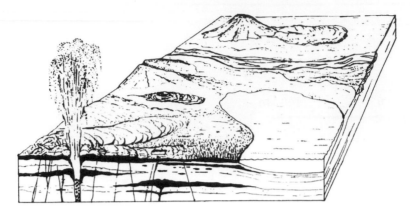

Fig. 9.4 Reconstruction of the environment at Rhynie, in 'Old Red Sandstone' times, showing volcanoes, a lake and geysir activity in a hydrothermal area. Reproduced by permission of the Royal Society of Edinburgh and N.H. trewin in *Transactions of the Royal Society Edinburgh: Earth Sciences*, volume 84 (1994, for 1993), pp 433–442.

Apart from the land flora and the small creatures that were beginning to emerge onto land, the early Old Red Sandstone lakes and wetlands provided habitats for rapidly evolving fish. From the jawless fish (distant relatives of modern lampreys and hagfish), first known from Ordovician fossils, the late Silurian and early Devonian Periods saw evolution of the jawed and heavily armoured placoderms. Thereafter there was rapid diversification to such groups as the early bony fish (ancestral to most modern fish) and, most importantly for us, lobefish. In Chapter 7 it was noted that early Carboniferous vertebrates had acquired lungs for terrestrial respiration and limbs for out-of-water propulsion. The ancestral line from these leads back to the Devonian lobefish. Contemporary with these fish populations were assorted invertebrate creatures. Notable among these were the eurypterids, an arthropod group reminiscent of scorpions, with sizes of up to 2 m. In contrast to these lowland habitats, the upland regions would have been rocky deserts that lacked both deep-rooted vegetation and soils that could modulate the rate of run-off of water and thereby slow down erosion.

Shetland and Orkney

The oldest volcanoes described in the previous chapter were those active at the start of the Carboniferous Period in the Scottish borderlands. Prior to these the youngest Devonian volcanoes were some whose record is found in the Shetlands and Orkneys. In these islands the volcanic rocks were produced by eruptions about 380 million years ago within, or on the borders of, a vast lake known as the Orcadian Lake. This covered much of what is now Caithness, Orkney and Shetland and came into being at around 394 Myr in middle- to late-Devonian times.

Outcrops of lavas, ashes and minor intrusions occur on the western side of Shetland at Esha Ness and the Walls Peninsula, and also on the island of Papa Stour. At Esha Ness some of the pyroclastic deposits represent ash-cones interbedded with lavas and sedimentary rocks, and the volcanic rocks range from mafic to salic varieties. Salic ignimbrites are present on both Esha Ness and the Walls Peninsula. The presence of ash-cones and extensive ignimbrites has prompted the suggestion that one or more calderas may originally have been involved. The Papa Stour occurrences are of particular interest in containing two very thick rhyolite units that are probably lava flows rather than ignimbrites. Among the scattered vents and small occurrences of lava on Orkney, the basalt lavas at the top of the succession on Hoy are of interest in being compositionally akin to the Carboniferous lavas erupted in the south of Scotland some thirty to forty million years later. Like most of the Carboniferous lavas they are relatively poor in silica but rich in the alkali metals.

Volcanism in the Borders and Southern Uplands

Not so very long (geologically speaking) before the volcanism in Orkney and Shetland, what appears to have been a single great volcano grew in the south-east of Scotland, dating from about 395 Myr. It straddled the present border between England and Scotland and thus lay across the 'suture zone' that marked the end of the Iapetus Ocean (*Fig. 2.1*). It is the products of this volcano, occupying some 500 km², that form the high moorland massif of the Cheviot. A thick basal unit of fragmental rocks is overlain by rhyolite lavas and these are followed by andesite and dacite lavas rather than basalts. Consequently the whole volcano was relatively rich in silica. Late-stage magmas, rising high in the volcanic pile, crystallised to three (crudely concentric) intrusions. The earliest of these, with a composition similar to that of the andesite lavas, formed a coarse dioritic body. The next two were distinctly more siliceous, forming granitic rocks. These major intrusions were associated with a swarm of dykes radiating out into the surrounding lavas. The magmas responsible for the dioritic to granitic core marked the final act at the Cheviot and, since they penetrated the preceding extrusive products, it is probable that they also had a surface expression. If so, the finale may have involved violent eruption of rhyolitic ash-flows, followed by emission of a viscous, de-gassed rhyolite dome.

Although the present outcrop area of the Cheviot lavas is some 40 km across, it is clear that much of the volcanic sequence lies beneath younger Carboniferous strata. We may consequently imagine that the Cheviot volcano was a very large edifice, perhaps around 60 km in diameter at its base and probably rising to a height of 2 or 3 km. The Cheviot volcano remains something of an enigma in that its origin cannot readily be assigned to any particular tectonic setting. Some 25 km to the north of the Cheviot there are excellent coastal exposures of somewhat older (c.410 Myr) basaltic lavas at St Abb's Head and Eyemouth. It is not known from which centres these were erupted.

The Midland Valley and the Grampian Highlands

Pre-dating all of the above, extensive volcanism had affected a broad tract of Scotland from the northern margin of the Southern Uplands to the Great Glen Fault. The lava sequences from this date from 425 to 412 Myr, although related intrusive activity suggest that eruptions probably persisted to about 395 Myr (i.e. essentially contemporary with the Cheviot volcano) and pebbles in some of the Old Red Sandstone conglomerates indicate older volcanic sources that are no longer to be found.

Despite varying widely in composition, all of the lavas of the Midland Valley and Grampian Highlands have chemical characteristics that point to their generation above a subducting plate of oceanic lithosphere. So, for the first time as we go back in the Scottish historical record, we encounter clear evidence for supra-subduction volcanoes. As outlined in Chapters 1 and 3, the descent of old, cold oceanic lithosphere for recycling in the mantle is typically attended by the growth of volcanoes behind the zone (normally marked by a deep submarine trench) where the lithospheric plate begins its downward plunge. We are confronted at this point with a dilemma. Whereas the closure of the Iapetus Ocean is thought to have been complete when the component continental pieces of Laurussia had finally docked, we find that magmas with subduction-related chemical signatures were still being generated up to 25 million years later. It would seem that we have to accept that, long after Laurentia, Avalonia and Baltica had locked together, some degree of subduction was still going on. This is inferred to have involved oceanic lithosphere plunging at a gentle angle beneath what is now the Midland Valley and the Grampian Highlands.

Analyses across modern volcanic arcs show distinct changes in their chemical make-up. Thus if, for example, one examines basalt lavas collected in a traverse (usually over distances of one or two hundred kilometres) perpendicular to an arc, the content of potassium increases as one progresses away from the zone where subduction commences (see *Fig. 1.8*). Since the plate from which the magmas originate is sinking, it follows that the magmas have to ascend from ever greater depths to supply those volcanoes situated at increasing distances behind the trench. In other words there is a positive correlation between the potassium content of the mafic lavas and the depth from which they originated, close to the upper surface of the plate. The recognition of this relationship between potassium content and plate (or slab) depth has provided a useful tool for geologists studying ancient arc volcanoes to find out firstly, in which direction the oceanic lithosphere was descending and secondly to make tentative estimates regarding its depth at any locality. It was from such analyses of the Silurian–Devonian volcanic rocks that the realisation came that a subducting slab of Iapetus Ocean floor must have been going down in a northwesterly direction beneath much of Scotland.

The Midland Valley

In the introduction to this chapter it was explained that the Midland Valley as we see it now almost certainly represents only a part of the crust that must formerly have existed between the Southern Uplands and the Grampian Highland terrane during late Silurian and much of the Devonian Periods. At this time the Midland Valley was a region of volcanic hills separating adjacent lowlands (*Fig. 9.3b*). Ephemeral lakes in the latter probably developed where river waters were impounded by lavas or mudflows. The rivers draining from both sides of the Midland Valley deposited huge quantities of boulders, pebbles and sands into two main subsiding troughs and, as a result of intermittent eruptions from local volcanoes, these sediments became intricately interleaved with the lavas. Not only was the high ground adjacent to the Midland Valley contributing sediment but there is evidence that the volcanoes themselves were being vigorously eroded in the interludes between one eruption and the next. Although the lavas commonly possess red surfaces due to oxidation, deeply weathered 'boles' (such as typify much of the Palaeogene basalts and, for example, the Carboniferous 'Passage Group' lavas) are rare. Their scarcity could merely reflect the prevalent dry climate and lack of plant-root penetration, but may also indicate the shortness of time before the lavas were covered up, either by the next lava or by the deposition of sediment. Study of the fluviatile sediments in the Midland Valley suggests not only that they were supplied from eroding mountains to the north and south but that a ridge of higher ground may have lain more or less axially, embracing what is now

Fife and the upper Forth valley and extending WSW towards the Clyde Estuary. Chains of volcanoes may have been situated on either side of this, as shown in the palaeo-geographic map (*Fig. 9.3b*).

The landscape involved numerous relatively small volcanoes built up from lenticular alternations of laterally impersistent lavas, with eruptions taking place over a span of some 13 million years from around 425 to 412 Myr. The small angular differences between the lava flows and accompanying sedimentary strata point to a general absence of steep cones and suggest that many lavas were the products of fissure eruptions. The lavas were typically of *aa* type, with rough jagged surfaces. Some spilled over dry land but many, for example in Ayrshire and Angus, encountered wet sands or muds, perhaps on lake floors. In such cases, instead of covering the surface, the magma frequently burrowed into the waterlogged sediment forming sills, often with pillowed margins and considerable admixture of detrital material.

After subsequent folding and faulting, the volcanic products were largely buried beneath great thicknesses of younger, mainly late Devonian and Carboniferous strata, thereby severely handicapping appreciation of the full picture. As a result, it is generally not possible to pinpoint where the individual volcanoes were situated. The best guesses have to come from identifying those localities where the lava piles are thickest or, alternatively, from shallow-level intrusions whose ages, forms and compositions suggest that they are volcanic roots. For example, a short distance south of Darvel in Ayrshire there is a short NE–SW chain of granitic to dioritic intrusions that may qualify as sub-volcanic. The largest of these, the Distinkhorn intrusion, about 5 x 2 km across, has been dated at around 413 Myr. Each of these early Devonian intrusions is surrounded by a broad zone of thoroughly baked sedimentary rocks. The magmas obviously attained very shallow crustal levels and, as in the case of the Cheviot granite magma, they could well have fed surface volcanoes. Given their relatively high silica contents, we are left to speculate whether such magmas also gave rise to explosive pyroclastic rhyolitic and dacitic eruptions, whereas the non-erupted portions, congealing in the conduits beneath a volcanic capping, crystallised to the coarse-grained rocks that are now left to our view.

Between the valleys of the Firth and Earn, volcanic rocks emerge on the north side of the fault that forms the great escarpment of the Ochil Hills, which reveals a great thickness (more than 2.4 km) of andesitic, dacitic and rhyolitic lavas and pyroclastic deposits (*Fig. 9.5*). Whereas pyroclastic deposits are generally insignificant in the 'Old Red Sandstone' volcanoes of the Midland Valley, they form a dominant part of the succession in the western Ochils. Some are very coarse, as indicated by Geikie's (1897)

Fig. 9.5 View of the Ochil escarpment from near Stirling.

observation that some blocks are 'as large as a crofter's cottage'. Although, as has been stated, it is generally impossible to be precise about the locations of the individual volcanoes, the observation that the thickness and coarseness of the pyroclastic rocks increase southwards towards the Ochil Fault suggests that the main eruptive centre lay south of the fault. If so, it now lies concealed by a huge thickness of younger sedimentary strata, beneath the Alva and Tillicoultry districts. Other possible eruptive centres could be marked by the four sub-cylindrical intrusions of dioritic rocks north of Tillicoultry, where the surrounding rocks have been intensively baked ('thermally metamorphosed') to a distance of some hundreds of metres out from the intrusions. If the heat was derived from large volumes of magma flowing up through the pipe-like conduits, these intrusions could represent the root zones beneath surface cones. The argument for these being sub-volcanic intrusions is similar to that outlined above for the Distinkhorn and its neighbouring intrusions.

The volcanic outcrops extend north-east from the western Ochils along the southern side of the Firth of Tay, as well as to the north of the firth, beyond Perth into the Sidlaw Hills. Lava sequences, intricately interleaved with the sandstones and conglomerates, can be followed still farther north-east towards the coast and are well exposed near Crawton and to the south of Montrose. Some were probably erupted from a volcanic centre now out to sea, beyond Montrose. To the east and south of Perth the lavas form the

Fig. 9.6 A view of Turnhouse Hill in the Pentlands, Midlothian. The terraces in the background, dipping from upper right to lower left, reflect individual basalt lava flows.

hills of Kinnoul and Moncreiffe. The motorist is afforded fleeting glimpses of the lavas in the deep road-cuts on the M9 between Perth and Bridge of Earn.

On the southern side of the Midland Valley Siluro-Devonian lavas outcrop in discontinuous lenses close by and roughly parallel to the Southern Upland Fault. They can be traced from the Carrick Hills SW of Ayr, north-eastwards towards Edinburgh. In the Tinto Hills there is a lenticular intrusion of rhyolite. This brick-red rock is one widely used as ornamental chippings for road and drive surfaces. Just south of Edinburgh, in the Pentland, Braid and Blackford hills, the succession is at its most varied (ranging from basaltic to rhyolitic) and thickest (at about 1,800 m). *Fig. 9.6* shows a view in the Pentland hills where weathering of the lavas, despite their very poor exposure, presents a series of dipping terraces. From thickenings of the volcanic pile the Pentland lavas are considered to have been emitted from three main centres. It has been suggested that one or more eruptive centres lay near the north end of the Pentlands, where pyroclastic deposits are well developed. Again, much of the critical evidence is buried beneath younger rock formations. The abundance of silica-rich products (dacites and rhyolites) suggests that an underlying magma chamber was large enough to permit extensive operation of fractional crystallization and crustal melting.

It was noted in Chapter 5 that the original gas-bubbles or vesicles in the Palaeogene lavas of the Hebrides have generally been filled with white or colourless minerals of the zeolite family, deposited from hydrothermal fluids during post-eruptive cooling stages. In the Siluro-Devonian lavas considered here, the watery solutions pulsing through the lavas are deduced to have been rather hotter (probably 300–400° C), too high for the formation of zeolite. Silica, in the extremely fine-grained form of agate, is the typical vesicle filler rather than zeolites. The agate was deposited in successive concentric shells, from the outside of the vesicles inwards. Subtle changes in temperature and water chemistry led to the agate layers having slightly differing colours. The agate-filled vesicles or 'amygdales' are extremely tough, insoluble and generally resistant materials that tend to survive even after the surrounding lava has crumbled and eroded. Consequently quite fine examples of banded agates may be picked up on the beaches, for example near Montrose, where they have been weathered out from their host lavas. The 'Scotch pebbles' of the Galston area, Ayrshire, are similar materials. Jasper, fine-grained silica stained red with haematite, may also be found in amygdales, as may mauve amethystine quartz.

The Grampian Highlands

The Siluro-Devonian conglomerates in the Midland Valley become thicker, coarser and abundantly rich in volcanic pebbles as traced towards the Highland Boundary Fault. While conglomerates rich in volcanic pebbles are abundant on Arran some of the most spectacular volcanic conglomerates are to be seen in Northern Ireland, in and around the caves at Cushendun, just south-east of the extrapolation of the Highland Boundary Fault south-west from Arran and Kintyre. Such deposits leave us in no doubt that contemporary volcanic centres lay to the north of the Highland line.

Although surface products of the Highland volcanoes have largely been eroded away during later periods of uplift, there is widespread evidence for their former presence. Small areas floored mainly by basalt and andesite lavas resting on deformed Caledonian rocks occur close to the Highland Boundary Fault in the Forest of Alyth and near Monzie. These lavas can be matched with those of Strathmore in the north-eastern sector of the Midland Valley. One very distinctive volcanic unit is a dacitic ignimbrite (welded ash-flow), up to 40 m thick, that provides evidence of an extensive 'glowing cloud' eruption that swept across the region. This ignimbrite occurs not only in the Blairgowrie area in the Highlands but also across the Highland Boundary Fault to overlie the narrow package of early Palaeozoic rocks, known as the Highland Border complex, which lies along the fault.

Lessons to be gleaned from these observations are that, at this stage, the Highland Boundary Fault did not present a topographic barrier and lavas and ash-flows could spill across it. Furthermore it proves that by this date the rocks of the Grampian Highlands, the Highland Border complex and the Midland Valley were all firmly locked together as a single coherent unit.

In north-eastern Scotland the small down-faulted sliver containing lavas and ashes near Rhynie, Aberdeenshire, has already been alluded to on account of its associated fossil-bearing cherts. Other scraps of Siluro-Devonian lavas in NE Scotland have survived erosion near Buckie and west of Cabrach, while volcanic pebbles occur in conglomerates on the Black Isle. The volcanoes from which these lavas came are again unknown: some may lie east of the present coast, although dioritic to granitic intrusions south-west of Lochnagar could possibly mark such centres.

Siluro-Devonian extrusive and intrusive rocks of Argyll

A conspicuous belt of igneous rocks, roughly 30 km wide, lies south-east of the Great Glen Fault, which it approximately parallels. The rocks, comprising lavas, dykes and major granitic intrusions (plutons), resulted from magmatism affecting the Grampian terrane between 425 and 395 million years ago. By this time orogenic deformation and metamorphism were long finished and the Highlands had already experienced intense erosion. The NE–SW trend of the belt was dictated by the overall 'grain' of the older rocks (folds, faults etc.), that had long since been imposed by the Caledonian Orogeny. South-west of this belt an extensive outcrop of lavas constitutes the Lorne Plateau and represents by far the largest tract of Siluro-Devonian volcanic rocks north of the Highland Boundary Fault. The lavas extend from Loch Melfort in the south-west to Loch Creran in the north-east and were erupted approximately 420 to 415 million years ago, i.e. contemporaneously with much of the Midland Valley activity.

The Lorne lavas are truncated to the north-east by an annular suite of dioritic and granitic rocks, forming the Etive Complex, which includes Ben Cruachan and Ben Starav. Still further to the north-east is another, older mass of granitic rocks, underlying Rannoch Moor and, between this and the Etive Complex is the Glen Coe area, largely composed of volcanic rocks and intrusive granites. Cutting through almost all of the above is a major NE–SW-trending dyke swarm (the Etive dyke swarm; *Fig. 9.7*). The words 'almost all' refer to the fact that some of the youngest intrusions of the Etive Complex may post-date intrusion of the swarm.

Clearly all the magmas involved in this Argyll belt either reached the surface or attained shallow levels in the crust. Their congruence in space

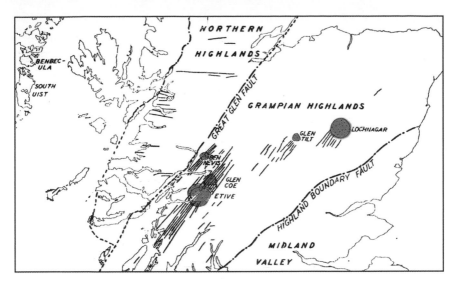

Fig. 9.7 The distribution of the principal Silurian-Devonian dyke swarms in the Highlands and their relationship to some of the larger igneous centres, including those of Ben Nevis, Etive and Glen Coe. Other NE-SW dykes shown are those associated with the Glen Tilt and Lochnagar granite intrusions. *(After J.E. Richey, 1939)*

and time implies that their genesis had much in common and all, in some manner, were related to the subduction of the Iapetus oceanic plate, which had reached its deepest levels in its transit across and beneath 'Scotland'. The magmas arose as a result of faulting and fissuring of a mechanically weak tract of the lithosphere that was responding to the same tectonic forces that caused huge left-lateral movements along the Great Glen Fault. I will make the further assumption that the Etive complex, and perhaps also the Rannoch Moor magmas, had surface expression as major volcanic structures and that the Etive dyke swarm likewise gave rise to volcanism.

Between these assorted igneous rocks and the Great Glen Fault there is another, subordinate NE–SW-trending belt, involving coarse-grained 'plutonic' rocks, lavas and volcanic rocks. This includes the Ballachulish granite, the Ben Nevis intrusive complex and a swarm of NE–SW-trending dykes, smaller but otherwise comparable to those of the Etive swarm. The probability is that the rocks we see in the Ben Nevis complex also crystallised deep beneath a central-type volcano.

The Lorne lavas

The Lorne lava succession is up to 800 m thick, with the individual flows up to 30 m. The outcrop area is approximately 300 km² but there can be

Fig. 9.8 Scenery due to the Lorne lavas near Easdale, south of Oban.

little doubt that what we see is only a portion of the area once occupied. Lavas which escaped erosional oblivion by down-faulting at Ben Nevis, some 20 km north-east of the main outcrop, were possibly part of the Lorne lava field. Others, sandwiched between two of the large intrusions in the Etive Complex, are more certainly relic portions of the Lorne sequence and some authorities have correlated andesitic units (flows or shallow sills?) at Glen Coe with those of the Lorne. Evidence that the lavas spread westwards onto Mull is provided by an outcrop in the south-east of the island, just south of the Great Glen Fault. Lava pebbles occur in Siluro-Devonian conglomerates in southern Kintyre and there is a single lava flow in Arran, interstratified with sandstones and conglomerates to the south of the northern granite. Whether or not these are indications of a much further southward extension of the Lorne Plateau lavas is an open question. The lavas are well exposed, e.g. north and south of Oban (*Fig. 9.8*), but are seen to best advantage on the island of Kerrera, particularly along its east and south-east coasts, where some of the surface features of the lava are excellently preserved. Typically these show the broken-up, slaggy tops characteristic of *aa* flows. Many appear to have erupted onto or into wet sands or gravels, and to have been affected by the consequent explosive generation of steam. Others appear to have over-run dry land and show the corded surfaces of *pahoehoe*-type flows.

Over much of their outcrop the lavas sit directly over rocks that had been folded and metamorphosed in the Caledonian Orogeny before being uplifted and deeply eroded. Near Oban, however, there is a thick basal sequence of conglomerate formed largely from volcanic pebbles, suggesting that one or more volcanoes had already been reduced by erosion before the first of the surviving lavas flowed across the area. The scenery at that time is likely to have been one of semi-desert low relief broken by volcanic hills. These, with rather low-angled flanks, were probably elongate NE–SW, in rough parallelism to the overall trend of the Caledonian rocks, and the underlying feeders are likely to have been dykes of the same orientation. Despite the fact that the Caledonian mountains had already been extensively worn down, they were still capable of providing coarse alluvium for rivers draining into the area and the lavas are locally interbedded with sands and conglomerates deposited in the quiescent interludes. That fish swam in the shallow waters is confirmed by the presence of fossils (*Cephalaspis lornensis*) in sandstones near the base of the sequence. Although the succession has a thickness approaching a kilometre, the surface of the lava field is unlikely ever to have been much above sea-level. Consequently it is reasonable to infer that the substrate was continually subsiding in order to accumulate so great a thickness of strata. After its formation the Lorne lava sequence was deformed into a shallow south-west dipping downfold.

The lower part of the succession is of basaltic and andesitic lavas but the upper part consists of somewhat more evolved compositions, lacking basalts but with andesites and salic lavas as well as two ignimbrites. Chemically the Lorne lavas are interesting in being richer in potassium than their counterparts further to the south and east of Scotland. The correlation between potassium content and depth to the subducted slab beneath modern supra-subduction volcanoes has already been discussed. Although it is not possible to attribute specific depths to the top of the inferred slab of Iapetus lithosphere, the change in chemistry shown by the Siluro-Devonian lavas across Scotland is the best proof that the slab dipped to increasingly greater depths as traced north-west towards the Lorne Plateau.

Glen Coe

Glen Coe, some 25 km to the north-east of the Lorne lavas (*Fig. 9.9*) has played an important role in the development of volcanological concepts. The realisation that caldera-forming events could be discerned in ancient eroded volcanic roots arose from mapping by the Geological Survey in the western Grampian Highlands nearly a century ago. The results of this

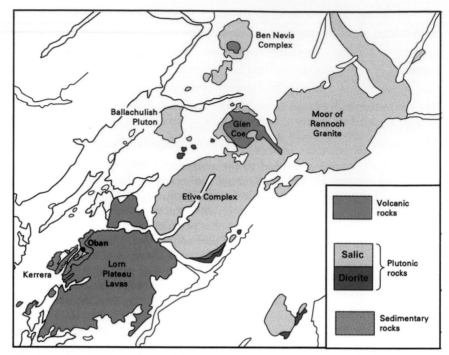

Fig. 9.9 Map showing relationships of the Lorne lavas, the Etive Complex, Glencoe, and the Moor of Rannoch granite. *(After British Geological Survey in British Regional Geology, The Grampian Highlands, 4th Edition, D. Stephenson et al., eds., 1995)*

inspired work, published in 1909 by C. T. Clough, H. B. Maufe and E. B. Bailey, gave the world its first detailed description and interpretation of a deeply dissected caldera volcano, now known to date from around 421 Myr. The mapping of Glen Coe and later of Ben Nevis led these three to postulate that sub-cylindrical masses of rock had subsided, piston-like, within ring-faults, producing calderas at the surface. The original Geological Survey investigations have been greatly amplified and elaborated by other geologists during the past forty years and the result is that we have now a very detailed account of the history of the Glen Coe volcano. This, in barest essence, turns out to be a story of a ding-dong battle between hot magma and cold water.

As we have seen in some of the Palaeogene examples (Chapter 6), down-faulting of masses of rock that were originally close to the surface within a ring-fault allows them to be preferentially conserved for posterity while their counterparts outside the fault are left exposed to the elements and often totally destroyed. This was the case at Glen Coe and, as we shall see, also at Ben Nevis. At Glen Coe the original survey concluded that subsidence occurred within a ring-fault bounding an ovoid cauldron (or caldera) some 14 km along its NW–SE axis and 8 km across. However, more recent investigations have shown that the formation of the caldera was not a single event involving the simple subsidence of a cylindrical plug, but was incremental, involving numerous fault blocks. NW–SE-trending faults bound a central subsiding basin (or 'graben') which is dissected by several NE–SW-trending cross-faults. It has been demonstrated that there was a multiplicity of movements involving different fault blocks at different times, so the generation of the structure was highly complex. Not only was there an intimate linkage between fault movements and eruptions, but erosion and consequent transport and deposition of sediment also played an integral part in the evolution of the volcano.

A sequence of strata sank within the volcano and granitic magma from an underlying magma chamber squeezed up the ring-fault to form a partial ring-dyke. Bidean nam Bian (*Fig. 9.10*) lies inside this fault and its cliffs expose the full thickness (over 1 km) of the subsided lavas, pyroclastic rocks and intra-volcanic alluvial sediments that had accumulated within the caldera. Much of the succession is also well exposed in Buchaille Etive Beag. The lowest igneous units are sheets of andesite and related rocks. These, initially thought to have been lavas, are now suspected to have been sills intruded into wet sediments.

Throughout the evolution of the volcano a river flowed north-westwards through the graben, carrying sediment towards the Great Glen. Of the five major ignimbrite eruptions within the Glen Coe caldera, the first three, with a total thickness of around 300 m, have been documented in detail. Each in turn commenced explosively as the rising magma encountered surface waters. The opening of fissures led to vigorous fountaining of gas and rhyolitic pumice: highly fluid ash-flows condensed from these fountains, flowing out to be trapped within the steep caldera walls and compacting as ignimbrite sheets. The surfaces of the ash-flows appear to have been essentially horizontal. Sills, up to 100 m thick, composed of poorly mixed andesite and rhyolite magma were injected into this ignimbrite succession in response to intermittent subsidences of the caldera floor. Whereas salic magmas, erupted as ignimbrites, were prominent in the later history of the volcano, the evidence of magma-mixing in some of the sills is an eloquent reminder that the higher-temperature, more primitive andesite magmas were still present at depth. In the intervals between eruptions the higher ground was eroding while river or lake sediments were being deposited. The first three ignimbrite eruptions are estimated to have taken place over about half a million years, during which descent of the volcano-graben

Fig. 9.10 View of Glen Coe showing the major ignimbrite units that form the massive terraced features in Bidean nam Bian.

floor continued at a rate of over 500 m per million years. Although intermittent, the individual subsidence events were dramatic and synchronous with the eruptions. While the ignimbrites are now confined within the collapse structure, it is likely that some also spread over the surrounding landscape and it has been suggested that the thin ignimbrites in the Lorne succession originated from the Glen Coe volcano. The compositions of the Glen Coe volcanic rocks have features relating them to those of the Lorne lavas but there are chemical differences, and relationships between the Glen Coe and Lorne andesites remain debatable.

The younger granites and their postulated association with large caldera volcanoes

Before continuing with a discussion of the Etive Complex, the Rannoch Moor granite and the Ben Nevis complex, each of which is largely made up of coarse-grained, so-called plutonic rocks (from their fanciful association with the underworld realms of Pluto), I will revert to the subject of salic magmas and crustal melting.

At the peak of metamorphism and deformation in the Caledonian Orogeny, temperatures in the deep crust of the Grampian Highlands would have reached values of around 750–850° C. At such high temperatures the rocks responded, not merely by recrystallising (metamorphosing), but by starting to melt to form salic magmas. We have encountered crustal melting in Chapter 6, with the generation of granitic magmas as a result of the heat arising from the ascent of basalt magmas coming from the underlying mantle. The large-scale crustal melting that occurred in association with the later stages of the orogeny was probably partly due to intrusion of mantle-melts at the base of the crust.

Hydrous fluids from the sinking Iapetus slab added potassium (and other components) to the overlying mantle wedge, modifying its composition and promoting its melting. The melts so formed, in addition to those produced from the oceanic crust itself as well as from the continental crust above, gave birth to the plutonic magmas. Although north-westward subduction of the Iapetus floor is thought to have commenced at about 460 Myr, there was a long delay of more than 30 million years before the emplacement of the Argyllshire magmas. Detailed structural analysis of the SW Grampian Highlands indicates that the deep crust between the Great Glen Fault and the Highland Boundary Fault was broken up into numerous coherent blocks that rotated against each other in response to movements on these two great dislocations. If one imagines the blocks as bricks in a layer that is being shuffled, it is easy to visualise points where the bricks are pressing hard against each other and other places where gaps open up. With this mental picture in mind, one can envisage there having been places where the deep crustal blocks were tending to separate, leaving extensional spaces between them. Rising melts took advantage of these to give rise to the big granitic plutons.

Great granitic plutons are abundant in north-east Scotland and underlie, for example, the Cairngorm Mountains. More than a dozen of these granite plutons are spread discontinuously across the Grampian Highlands, eastwards from the Cairngorms to the coast near Aberdeen. Although the magmas for these intrusions rose high into the crust there is no evidence that they fed surface volcanism. Other Siluro-Devonian granitic plutons emplaced in the Caledonian mountains lie in a scattered chain lying to the NW of the Cairngorms, from the vicinity of Nairn in the north-east towards Oban in the south-west. Some, like the Ross of Mull and the Strontian plutons (in which the great Glen Sanda 'super-quarry' is worked) lie on the north side of the Great Glen Fault.

At this point it is worth recalling that magmas relatively rich in silica can give rise to very different sorts of rock according to circumstances. Slow crystallisation of the hydrous ('wet') magmas beneath an impermeable cap when loss of the water vapour is prevented, yields coarse-grained rocks like diorite and granite. Where the lid fails and high-pressure water can come out of solution, violent to cataclysmic eruptions ensue. Expelled diorite magma will congeal rapidly to fine-grained andesite and likewise, granite magma will yield rhyolite. So, one usually ends up with either a thoroughly coarse-grained intrusive rock or a very fine-grained (or, indeed, often glassy) extrusive rock. Some of the dioritic and granitic occurrences described below may well represent the deeper, bottled-up, magma fractions that 'stewed' with their 'juices' intact beneath great volcanic superstructures of previously erupted extrusive rocks (cf. *Fig. 3.10*). Although crustal melting had gone on throughout the Ordovician and Silurian periods, the bulk of the resulting magmas, including those found in the Etive, Rannoch and Ben Nevis plutons, are of late Silurian to early Devonian age, and are among the youngest in the Highlands. However, in contrast to the magmas that formed the Palaeogene granites (Chapter 6), the magmas under discussion are inferred to have possessed distinctly higher contents of dissolved water, derived from dehydration processes affecting the down-sinking oceanic plate far beneath them. The salic intrusive cores to some of the Palaeogene central-type volcanoes tend to be rocks of modest crystal size (1 to 2 mm) for which the terms microgranite or granophyre are generally appropriate. By contrast, the Siluro-Devonian 'plutons' are typically very much more coarsely crystalline with feldspars, quartz and other components several millimetres across. Whereas this may relate, to some extent, to our seeing deeper erosional levels, the larger crystal growth was primarily due to their having crystallised from more water-rich magmas than was the case for the Palaeogene magmas.

During the slow crystallisation of the more silica-rich magmas, progressive growth of water-free crystals (feldspars and quartz) led to a rise in the concentration of water dissolved in the late-stage magmas. Ultimately the increasing pressure on the roofing rocks resulted in their fracturing. The ensuing catastrophic eruptions would have been characterised by tall (plinian) columns of gas and ash reaching high into the stratosphere with consequent widespread fallout of pumice particles together with 'glowing-cloud' ash-flow avalanches of gas and suspended pumice particles cascading down the flanks. High-velocity laterally directed gas blasts would have given rise to 'surges' of the kind produced by Mount St Helens in Oregon in 1980 (*Fig. 4.7*). Evacuation of magma during these gigantic eruptions led to caldera collapse. However, it should be recalled that the volcanic centres under consideration were each intimately associated with the swarms of NE–SW-trending dykes, whose emplacement continued until late in the story. Accordingly it is possible that some of the pre-caldera magma chamber emptying was due to sideways withdrawal of magma along propagating dykes as crustal extension took place.

The Etive Complex

The intrusions constituting the Etive Complex, which date from around 400 Myr, lie roughly concentrically one within another (*Fig. 9.11*). Erosion has revealed them as they were at a depth possibly as shallow as 3 km, and not more than 6 km. Ben Starav is carved out of the youngest pair of these intrusions, while Ben Cruachan consists of earlier units that form an outer shell around the Starav intrusions. Although the Ben Starav magmas were rich in silica, crystallising to form granites, the preceding Ben Cruachan magmas were somewhat more deficient in silica, the earliest crystallising as diorite. This pattern probably denotes increasing contributions of crustal melts as time went by. The intrusions forming Ben Cruachan are cut by the NE-trending dykes but the two younger (Ben Starav) plutons are dyke-free and may represent the culminating magmatic activity in the whole NE–SW igneous Argyll province (*Fig. 9.11*).

Between two of the earliest components of the Cruachan complex lies a mass of lavas with some interbedded sediments. It forms a vertical 'screen',

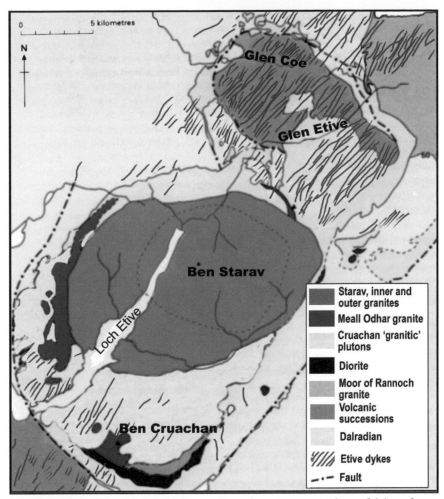

Fig. 9.11 Map of the Etive and Glencoe Complexes. Large numbers of dykes of the Etive Dyke Swarm transect Glen Coe. Fewer cut the earlier units of the Etive Complex while only a very few late-stage dykes intruded the youngest granitic units of Ben Starav. *(After G.S. Johnson, 1966)*

crescentic in plan, about a kilometre broad and some 8 km long. The lavas are clearly derived from the Lorne sequence which has otherwise been eroded away from this area, saved from a similar fate by down-faulting during a successive series of sub-cylindrical collapses as the Ben Cruachan magmas were emplaced. Each of these collapses was probably accompanied by surface eruptions and caldera-formation.

The Rannoch Moor granite was already undergoing erosion during the time when the Glen Coe volcano was active. Since the Ben Cruachan, and still younger Ben Starav, intrusions post-date the activity at Glen Coe, it appears that there was a very generalised migration of magmatic activity from north-east to south-west along the rifted zone denoted by the Etive dyke swarm. It may be recalled that similar shifts of focus are not uncommon in volcanic systems and we have seen examples, e.g. in the evolution of the Mull volcano (Chapter 6). The Rannoch Moor intrusion and the Ben Cruachan–Starav intrusive complex measure up to around 25 km in diameter and are much larger than the Glen Coe volcanic complex. If, as is suggested here, they represent the deeply eroded anatomy of siliceous volcanoes in early Devonian times these would have been huge structures, with basal diameters of at least 70 km.

Ben Nevis

After completing his work on Glen Coe, Maufe went on to study Ben Nevis, where he recognised a similar style of intrusive events (*Fig. 9.11*). If, as is supposed here, a volcano overlay Ben Nevis, erosion has wiped the volcano itself from the surface of the Earth and evidence for its former exist-ence has to be deduced from the remaining clues. The present mountain is composed of coarse dioritic and granitic rocks, together with a central plug-like mass of volcanic and sedimentary rocks that forms the summit area. The Ben Nevis complex was very roughly contemporaneous with that at Glen Coe. The whole Ben Nevis igneous complex is sub-circular, about 6–7 km in diameter with the plug in the middle being about 2 km across, the whole ensemble thus being considerably smaller than that at Glen Coe.

Ben Nevis represents a sub-volcanic igneous complex composed of several successive and partially concentric intrusions (*Fig. 9.13*). The first two consist of dioritic rocks while the two following are granitic, so that the pattern of development was comparable to that in the Etive complex. While each of these intrusions penetrated the metamorphosed Caledonian country-rocks, the youngest, at least, also invaded a cover of volcanic and sedimentary strata. As the youngest granitic magma neared the surface, a ring-fault was generated in the volcanic roof permitting a great cylindrical plug, embracing about 700 m thickness of the cover rocks, to sink into the magma. Although the volcanic rocks now form the summit of Ben Nevis, we must accept the evidence that they have been down-dropped hundreds of metres by this 'cauldron subsidence' mechanism to their present position. These rocks, which are well exposed in the north-facing cliffs of Ben Nevis, embrace a wide variety of lavas, from andesitic, dacitic to trachytic, interbedded with coarse fragmental volcanic rocks and finer sedimentary beds, and are probably a representative sequence of what was originally a regional succession. The fragmental volcanic materials include the fallout from explosive eruptions as

Fig. 9.12 View of Ben Nevis from the north-east.

well as the unsorted detritus of mud-flows. Fine-grained lake deposits are also present and some resemble those of modern 'playa-lakes', i.e. lakes subject to seasonal or periodic drying out. Other interleaved sedimentary rocks include conglomerates indicative of occasional flash floods. Since some of the pebbles are of metamorphic rocks from the older Caledonian formations, the implication is that the area had considerable relief at the time, with the metamorphic rocks outcropping in upland regions. In brief the strata composing the plug record a down-sinking basin filled with lavas and sediments that had existed prior to the intrusion of the Ben Nevis complex and its caldera. According to the changing circumstances, lavas, ash-flows, mud-flows, screes, lake and river deposits accumulated in this basin.

The Ben Nevis ring-faulting penetrated a thick sequence of strata that formed at the surface and accordingly the chances are good that it too reached surface-level to form a caldera with concomitant eruptive activity. The granite that we see surrounding the subsided central plug is coarse grained. Although from this it can be inferred that it crystallised slowly with retention of its volatile components, a precursor magma probably did manage to break through to the surface and erupt, losing its gas content in pyroclastic eruptions. Those eating their sandwiches on the summit of Ben Nevis might pause to consider that, had they been there a little more than 400 million years earlier, they would have been sitting beneath the caldera of a volcano, situated a short way south of the Equator.

The Etive and Ben Nevis dyke swarms

The Etive dyke swarm is about 20 km broad and extends for some 100 km from Jura to 15 km NE of Kinlochleven (*Fig. 9.7*), cutting the Lorne lavas, the Rannoch granite and all the components of the Glen Coe centre as well as the earlier units of the Etive Complex. Dyke widths are typically less than 10 m and often less than 3 m. In some sectors the swarm is very crowded. Thus of the 9 km length of the Glen Coe caldera, roughly 4 km of elongation is due to the great number of dykes of the Etive swarm. Since the swarm cuts the older units in the Etive Complex (*c.*401 Myr) but appears older than the late granite at Starav (*c.*396 Myr), it can be dated as roughly 400 million years old. Early members of the swarm are relatively salic but younger ones are somewhat more mafic, suggesting an increased contribution from mantle as opposed to lower crustal sources. Their emplacement involved not just simple pull-apart extension but some left-lateral displacement, doubtless linked to the left-lateral movements that were taking place along the nearby Great Glen Fault.

The Etive dykes reached such high levels in the crust that it is a reasonable inference that they too supplied surface volcanism. Although recent studies have suggested that some five individual volcanic centres may have been active, no extrusive products survived the later uplift and erosion of the Highlands. The Etive dykes could represent a rejuvenation of the earlier fissuring that had allowed the Lorne magmas access through the crust. In the Ben Nevis area, regional extension allowed dyke intrusion to continue after the emplacement of the early intrusions (*Fig. 9.13*). However, this came to a close and the latest granitic intrusion post-dates almost all of the dykes. Thus it would seem that the lithosphere beneath Argyll was still being stretched and sheared to allow the intrusion of dykes during the early growth stages of the Etive Complex, but that these tectonic forces were fully expended by the latest stages, when the Starav magmas came in. We have seen that the dykes transect the Moor of Rannoch pluton and all the Glen Coe rocks, as well as the earlier units of Ben Cruachan. Similarly, NE-trending dykes cut the early units of the Ben Nevis Complex, but not the later ones. If, as has been argued, such dyke swarms included feeder channels for surface eruptions, it is reasonable to consider that the former central-type volcanoes hypothesised for Rannoch, Glen Coe, Etive and Nevis were themselves split, with the formation of fissure volcanoes or fissure-aligned craters for part or all of their development. Central-type dacitic to rhyolitic eruptions may, however, have attended the culminating stages at Ben Nevis and Ben Starav.

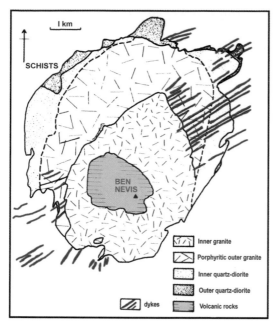

Fig. 9.13 Geological map of Ben Nevis. *(After Pankhurst and Sutherland in Igneous Rocks of the British Isles, D. S. Sutherland, ed., John Wiley & Sons, 1982)*

The Appinites

The appinites constitute an enigmatic suite of igneous rocks, ranging from ultramafic through mafic to salic, that is characterised by relatively high water and potassium contents. These components give rise to the presence of conspicuous black prisms of hornblende. The appinites are quite widespread in the Western Highlands but gain their name from their relative abundance in the Appin district, with the type examples from the southern shores of Loch Linnhe close to Ardsheal. The appinites typically occur in clusters of smallish pipes (50–500 m in diameter), some of which are ovoid in plan while others are irregular. They show a close spatial relationship to the late Siluro-Devonian plutons described above, occurring around them and sometimes within them. Some pre-date the plutons while others appear to be essentially contemporaneous with them. The intrusive appinites are often intimately associated with breccia pipes clogged with angular broken rock fragments, typically derived from the adjacent country-rocks. These pipes are thought to have formed from gases rising at high velocities in advance of the appinitic magmas and the presumption is that they gave rise to explosive eruptions at the surface. We may suppose these to have started with cones of rock debris blasted out by the gases. Whether these were ever succeeded by lavas or pyroclastic eruptions of the appinitic magmas remains a mystery.

Further speculations concerning the Argyll volcanism

Volcanoes characterised by extremely energetic eruptions, such as may have attended the postulated Argyll volcanoes, commonly have very low and seemingly innocuous profiles. While those under consideration may not have risen many hundreds of metres above the adjacent plains, they were likely to have been crowned by impressive calderas. In the case of the Etive and Ben Nevis complexes, these may have been nested calderas with diameters matching those of the essentially circular outlines of the plutons now seen on the geological map. The flanks of the volcanoes would have been heavily veneered with layers of air-fall pumice and the products of ash-flow eruptions, while lavas probably played a much less significant role within the extrusive products. These volcanoes have been much more deeply eroded than the roughly contemporaneous volcanoes of southern Scotland (e.g. the Cheviot), having experienced more profound subsequent uplift. Judging from the behaviour of comparable recent volcanoes, their eruptions would have been separated by spans of thousands or tens of thousands of years, but are likely to have been very voluminous and to have involved the violent expulsion of hundreds of cubic kilometres of magma. The eruptions themselves may have lasted anything from a few hours to a few months.

Although it is not possible to identify precisely analogous scenarios at the present time, some of the major volcanoes of the American cordilleras may offer analogies. One of these occurs in the down-faulted Owens Valley of eastern California, lying between the high mountains of the Sierra Nevada to the west and the White Mountains to the east. Here a major volcanic system, involving rhyolitic magmas, composes the area of Mono-Inyo and Long Valley. Another, more famous, salic volcano, situated some 800 km to the north-east across the Montana–Wyoming border, is that at Yellowstone. The long repose periods of these fearful volcanoes may be periodically interrupted when new accessions of mantle-born basaltic magma are injected into the crust, providing the thermal energy to re-activate large granitic magma chambers resident at higher levels in the crust.

At Yellowstone the volcanic products cover nearly 9,000 km^2 and one may imagine a comparable area to have been covered by air-fall and ash-flow pyroclasts following successive caldera-forming events at, for example, the Etive complex. Several times in the last 2 million years the magma chambers beneath the Yellowstone region have refilled with magma. Comparable time intervals, of the order of hundreds of thousands of years, probably elapsed between successive intrusions and accompanying caldera forming events in the volcanoes that may have existed in the south-western Grampian

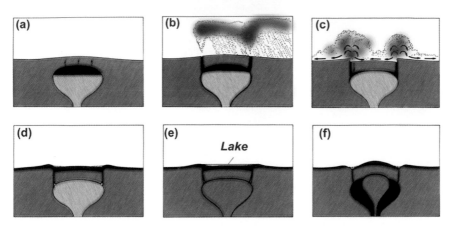

Fig. 9.14 Evolutionary stages of a salic 'mega-volcano', before, during and after a major eruption. *(After P. Francis, Scientific American Inc., 1983)*

Highlands during early Devonian times. It is the heat given off during the slow crystallisation of large magma bodies beneath Yellowstone that supports the hot springs, steam jets and geysers for which the region is celebrated. (Indeed it was the brightly coloured hydrothermally-altered rocks that gave Yellowstone its name.) On the basis of the scenarios postulated above, similar scenes (cf. *Fig. 5.19*) would have been enacted at Ben Nevis, Glen Coe etc. during their periods of magmatic repose. Although the American examples cited may be regarded, none too fancifully, as 'the sleeping dragons beneath the mountain', we can at least rest assured that those of the Scottish Highlands are indeed 'deader than the dodo'.

Fig. 9.14 shows, schematically, how these great volcanic systems operate. In (a) a shallow-level chamber fills with magma; in (b) ring-fracturing punctures the roof and pumice and ash are blown high into the atmosphere; in (c) the chamber empties and collapse and caldera formation ensue while ash-flow eruptions occur; (d) shows the low-relief profile of the volcano following a major eruption, with the caldera and surrounding region blanketed with pyroclastic products; in (e) a long period of repose occurs during which a lake develops in the caldera; and in (f) renewed ascent of magma, 'ballooning' in the low-pressure, near-surface environment, causes up-doming of the rocks above, creating a 'resurgent caldera'.

Chapter 10

Volcanoes and the Iapetus Ocean

The last chapter looked at volcanism affecting the Old Red Sandstone Continent that came into existence after the Iapetus Ocean had closed. We have seen that the Caledonian mountain belt was created as the fragments of the former continent were drawn into an oblique collision and that this process took place over many tens of millions of years. The geography during this time is likely to have been highly complex, probably similar to what we see today between Australia and south-east Asia, where two continental masses are closing on each other. In that part of the world there is an intricate pattern of volcanic island arcs embracing Indonesia, the Philippines and New Guinea, together with the intervening seas receiving the sediments brought down by the rivers from the adjacent emergent highlands. Much of this sediment comes from the erosion of volcanoes and can be categorised as volcaniclastic. Much of the region comprises deep trenches where ocean floor commences its descent (i.e. subduction) and volcanic arcs grow on the concave side of the trenches (cf. *Fig. 1.10*). Behind the arcs are the so-called back-arc basins, modern examples of which include the South China Sea, the Sea of Japan and the Java Sea. The structure of the lithosphere beneath the back-arc basins is also complicated, with some parts involving thinned continental crust but other parts with structures that roughly mimic that beneath the open ocean.

Although the oceanic lithosphere is very largely generated at the mid-ocean ridges (spreading centres or axes), a by no means insignificant proportion is composed of the products of intra-plate magmatism, unrelated to the plate boundaries. Products of the intra-plate activity include volcanoes that stay submerged as 'sea-mounts' as well as those that rise above sea-level as oceanic island volcanoes. In the Pacific the Hawaiian islands and those of Tahiti are celebrated examples of the latter. Comparable geographical and bathymetric features may be presumed to have been as significant in the early Palaeozoic as they are today.

In such a scenario of ocean closure, the geography would, in geological terms, have changed rapidly. Furthermore, it is necessary to bear in mind the generalised northward migration of the various pieces of continental lithosphere through time (Chapter 2) so that in the early Palaeozoic those eventually destined to participate in modern Scotland lay well to the south of the Equator. It is when one comes to consider the pre-Old Red Sandstone rocks of Scotland that interpretation of the geology presents great difficulties. Although vivid insights into aspects of local palaeogeography and palaeoenvironments can be gleaned from the rocks, there are serious problems in attempting large-scale syntheses. Unsurprisingly, opinions differ widely between investigators as to overall interpretations. I shall first consider some of the volcanic rocks from the Ordovician (i.e. roughly 488 to 443 Myr), erupted in association with the closure of Iapetus and, having done so, I will go further back in time to the volcanism believed to have attended the initiation of Rodinia's disintegration and the start of Iapetus.

The great bulk of the volcanic products considered in the earlier chapters were erupted on land. In contrast, almost all the lavas that we see in Scotland that pre-date the Caledonian Orogeny were erupted under water. Most of the eruptions happened on the sea floor, at depths of tens to hundreds of metres. We saw (Chapter 4) that sub-aerial, high-temperature basalt normally advances as *pahoehoe* flows, with toe-like rounded lobes at their front from which new lobes continually break out. The rounded forms are due to surface-tension forces acting on the molten material where it is in contact with air. It is such surface tension that causes, for example, rain-water to 'ball up' into rounded raindrops. The surface tension effects involving basalt magma and water are even more pronounced than those between basalt magma and air and the lavas typically advance in elongate tube-like protrusions. In section, a pile of these tubes or 'toes' can resemble pillows and are accordingly called pillow lavas. Some very shallow-water pillow lavas have already been encountered earlier with respect to the Palaeogene volcanism (Chapter 5).

Individual pillows are commonly up to half-a-metre or so in diameter. As the outer surface of each 'pillow' loses heat more rapidly than the interior, it usually congeals to a glassy rind a centimetre or so thick. While still hot these outer rinds behave as a deformable, leathery skin. The upper surfaces of the pillows will have an upwardly rounded (convex upward) form, while the bases of the pillows will settle in to occupy the cusp-like hollows between underlying pillow tops. In an accumulation of pillows their appearance in section may resemble that of stacked sand-bags, as shown in *Fig. 10.1*. Typically then, the products of the early Palaeozoic and late Precambrian volcanoes are almost invariably found as pillow lavas.

Fig. 10.1 Undeformed pillow lavas in Iceland, erupted within the past few million years. Knife 4cm long.

Ordovician volcanic rocks

Submarine lavas, pyroclastic deposits and associated intrusions of Ordovician age are not very abundant in Scotland, but are widely distributed throughout the Southern Uplands and on either side of the Midland Valley. Although there have been claims that some are relics of fully oceanic origin (i.e. produced at the Iapetus mid-ocean ridge) these have been disputed and it may be that what we find preserved are relics of back-arc basin floors. Although a very brief résumé of oceanic plate tectonics was presented in Chapter 1, here we must enter into a little more detail about the lithosphere beneath oceans and back-arc basins.

Intermittently, where parts of the lithosphere underlying the ocean or back-arc basin are too buoyant to subduct, they get thrust up against or over the margins of the converging continents. This process is called 'obduction'. Some of these obducted slices, produced during the closure of Iapetus in Ordovician times, can be found along the Norwegian coast while other superb examples occur in Newfoundland. Those considered here, in varying states of faulting and folding, have become part of the Scottish crust. The obducted slices can involve a very wide range of types of rock including pillow lavas (normally basaltic), together with intrusive rocks which are commonly doleritic dykes and masses of more slowly-cooled gabbro. Altered ultramafic rocks known as serpentinites are also normally associated and are interpreted as former mantle peridotites that have undergone low-temperature (< 500° C) hydration during obduction. In the course of hydration soft minerals of the serpentine group develop. These have a smooth and soapy feel to them and some serpentinite varieties ('soapstone') can be used for carving and turning. The name 'serpentinite' derives from the vaguely snake-like shiny and mottled appearance of the rocks. Some sedimentary rocks are often found together with these very diverse igneous rocks: hard, silica-rich rocks called cherts, largely formed from accumulations of siliceous marine micro-organisms called radiolarians, are prominent among these.

Serpentinites being a characteristic component of these obducted masses, the whole assemblage is called an ophiolite complex (from the Greek *ophi* meaning a serpent or snake). Ophiolite complexes are frequently crushed and deformed and their constituent rocks have usually experienced considerable changes in chemistry and mineralogy – i.e. have undergone some degree of low-temperature metamorphism. One might think of them as the geological equivalent of household waste that has gone through a garbage crusher. Nonetheless, huge amounts of geological history are potentially extractable from these unpromising assemblages. In Scotland the most notable ophiolitic assemblages constitute the northern islands in the Shetlands (Unst and Fetlar) and are also found on either side of the Midland Valley. Only the latter will be described here since the extrusive rocks are not preserved in the Shetland occurrences.

Before attempting any description of the Ordovician ophiolitic occurrences, I will try to explain how and why rocks as different as pillow basalts, dykes, gabbros and serpentinites came to be associated in the first place. In Chapter 1 it was explained that mantle peridotite beneath the mid-ocean ridges starts to melt as extensional thinning of the oceanic lithosphere causes the pressure on it to be relaxed. The basaltic melts so formed percolate up and accumulate to form magma chambers in the lower portions of the oceanic crust. Some of this magma cools slowly to form gabbroic rock. The crustal rocks above the magma chamber are relatively cool and, being under tension, fracture and allow further ascent of magma as dykes. The process is intermittent but repetitive and dyke after dyke is intruded, filling the potential void space generated as the oceanic plates on each side are pulled apart. Since the dykes are individually thin (usually less than 3 m), vertical and sub-parallel sheets, their repetitive emplacement creates

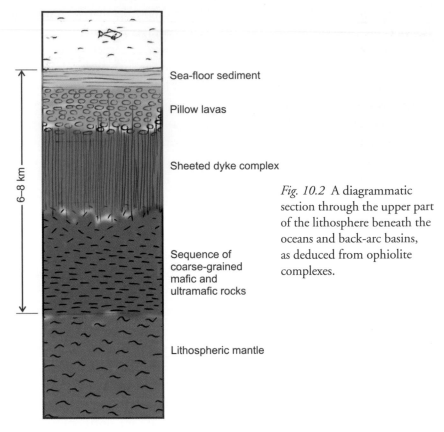

Sea-floor sediment

Pillow lavas

Sheeted dyke complex

Sequence of coarse-grained mafic and ultramafic rocks

Lithospheric mantle

6–8 km

Fig. 10.2 A diagrammatic section through the upper part of the lithosphere beneath the oceans and back-arc basins, as deduced from ophiolite complexes.

a sheeted complex with a vertical thickness of two or three kilometres, composed of virtually nothing but dykes. The youngest dyke components will be those directly over the spreading centre, with dykes becoming increasingly old as the distance away from the centre increases. The majority of the dykes will lose heat on the way up and crystallise before reaching the sea floor, but others get all the way and erupt their contribution of pillow lavas to the mid-ocean ridge. The combination of oceanic crust (about 6 to 8 km thick), together with a further thickness (several kilometres) of the used-up peridotite from which magma has been extracted, composes the inert oceanic lithosphere or lithospheric plate.

The peridotite shown at the base of the section in *Fig. 10.2* will be mantle rock from which the basaltic component has already been extracted. Overlying this will be the gabbros and associated mafic/ultramafic rocks crystallised from the expelled and ponded magma. This coarse-grained zone passes upwards, relatively abruptly into the sheeted dyke complex and, above this, forming the upper oceanic crust, are the pillow basalts, with or without some associated sedimentary rocks. This sequence of

Fig. 10.3 1969 eruption of Kavachi volcano, Soloman Islands, S.W. Pacific Ocean.

rock layers is thought also to reflect the general nature of the lithosphere beneath the back-arc basins as well as the 'open' oceans, and it is this overall assemblage of obducted rocks which contributes, in part or in whole, to the ophiolites. It should be stressed that the section shown in *Fig. 10.2* is very much an idealised one for there is very much variation between one ophiolite and another. Despite forty years of research into ophiolites around the world, as well as various studies of deep drill-cores from the oceans and back-arc basins, our knowledge of their lithospheric make-up remains at an early stage.

In addition to the intrusions and lavas that form at the spreading centre(s), the ocean floor and the back-arc basin floors will also have an irregular distribution of intra-plate volcanoes unconnected with the spreading centres. Although subordinate to the rocks formed along the mid-ocean ridges (spreading centres), these oceanic sea-mount and ocean island volcanoes do contribute a significant mass and volume to the oceanic crust. In addition to the igneous rocks resulting from spreading centres and intra-plate volcanoes, there will be others produced by island-arc volcanoes grown behind the trenches, above the subducting ocean floor plates (cf. *Fig. 1.8*). While the rocks we see were mostly, if not all, deposited or erupted on the sea floors, there is evidence that some of the intra-plate and arc volcanoes erupted subaerially. *Fig. 10.3*, an aerial photograph of

Kavachi Volcano in the Solomon Islands, shows a newly emergent island-arc volcano, with the surrounding waters contaminated by suspended particles which will be added to volcaniclastic sediments on the sea floor. Such a scene would have been commonplace in the supra-subduction volcanoes marginal to Iapetus.

The Highland Border Complex

Along the geologically complex zone adjacent to the Highland Boundary Fault, tectonic slices of rocks, with outcrops which are usually less than 1 km wide, are found as intermittent occurrences from Stonehaven in the NE, via Aberfoyle, Callander and the southern end of Loch Lomond, Bute, to northern Arran (*Fig. 10.4*). A continuation of this zone of faulted slices extends south and west to Co. Tyrone in Ireland. The ophiolitic rocks that compose them constitute the Highland Border Complex, sandwiched between the Grampian Highlands to the north and the Old Red Sandstone rocks that overlie them to the south. The complex embraces the wide variety of types of rock – including serpentinites, gabbros, dolerites and pillowed basalts – that modern geology recognises as implying some sort of oceanic derivation. Some sedimentary rocks, including black mudstones,

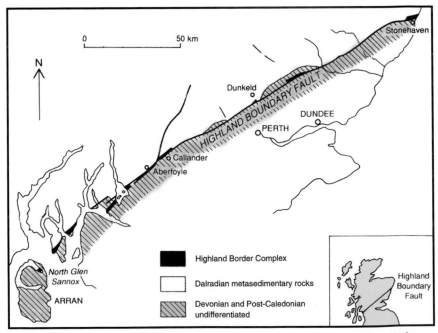

Fig. 10.4 Map showing the distribution of the Highland Border Complex. (*After P. E. Brown, 1991*)

Fig. 10.5 Pillow basalts in the Highland Border Complex, North Glen Sannox, Arran.

limestones and cherts, are also present. Although these ophiolitic assemblages now constitute narrow slices, there is consensus that the original terrane in which they formed was very much wider but was sheared, smeared and compressed against the continental basement underlying the Highlands during one or more collision events. The different rock units have been jammed against each other along steeply faulted boundaries, along some of which large lateral movements have been postulated. Some of the oldest rocks in the complex are of Precambrian to Cambrian provenance although the obduction of these ophiolitic assemblages took place during Ordovician times. Possibly as early as 430–420 million years ago (in the Silurian Period) the Midland Valley terrane, the oceanic or semi-oceanic Highland Border Complex and the Grampian terrane to the north were firmly welded together and, from this time on, the essential architecture of modern Scotland was established.

Unravelling the palaeogeography and the events that gave rise to the Highland Border Complex is a controversial task of immense complexity and we shall probably never have more than the most general idea of what was entailed. At a number of localities along the Highland Boundary Fault zone, the pillow lavas retain clearly recognisable forms (*Fig. 10.5*).

The Ballantrae Complex

In south-western Scotland, where the Southern Upland Fault zone passes off shore towards Northern Ireland, a remarkable suite of ophiolitic rocks outcrops over 75 km². It is well seen on the coast near Ballantrae, where serpentinised peridotites, gabbros, dykes (although, admittedly, not composing a sheeted complex) and pillow lavas occur, together with some cherts and black mudstones (*Fig. 10.6*). Much as in the Highland Border Complex, the Ballantrae Complex represents a disparate suite of rocks that has been forcefully amalgamated during the closure of Iapetus. It consists of a set of slices, brought into contact along NE–SW-trending faults, in which rocks from different depths in the crust and upper mantle are juxta-posed. Some of the sedimentary and volcanic rocks are of shallow-water origin, others are of deeper-water provenance while others, and especially the serpentinised peridotites, represent crustal or upper mantle rocks that were originally ten or more kilometres down. The complexity of the current geology in SE Asia–Australasia has already been alluded to and may be borne in mind when considering the Ballantrae situation, where rock suites from widely differing tectonic settings were brought into mutual contact.

As a consequence of their eventful tectonic history, the rocks of the Ballantrae Complex have been so extensively mangled, broken and hydrothermally altered that their interpretation remains a fertile field for academic dispute. Where stratified sequences of rock are recognisable, they have been dated partly by radiometric methods and partly on the basis of their fossil content. The fossils, which are few and far between, consist of the horny skeletons of an extinct group of planktonic, colonial organisms called graptolites which thrived in the early Palaeozoic oceans. As a group, graptolites are a god-send to geologists, firstly because they evolved fast and the different forms are very specific for different stages of the Lower Palaeozoic and secondly because they were free-floating components of the plankton with a very wide global distribution.

Although the details will, no doubt, be argued over for decades to come, there is general consensus that most of the rocks were generated in early Ordovician times and were obducted onto continental crust relatively soon after their formation. Within the Ballantrae Complex the lavas can be subdivided into various categories, erupted from volcanoes in differ-ing tectonic environments. Some, a minority, are regarded as having been formed at mid-ocean Iapetus constructional plate boundaries while others appear to have been formed in back-arc basins. Some were from island-

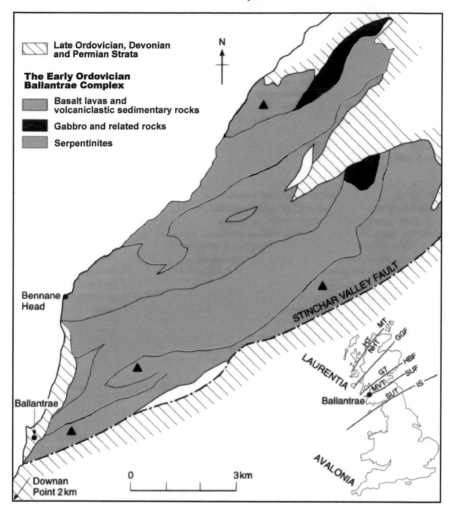

Fig. 10.6 Geological map of the Ballantrae Complex, Ayrshire. *(After G. J. H. Oliver et al., 2003)*

arc volcanoes above subducting plates and yet others are deduced to have been parts of intra-plate volcanoes. Evidence from igneous pebbles in some of the sedimentary rocks indicates that some came from subaerial cones. All were crushed together in the climax of an event which may well have involved collision of an oceanic island arc with the margin of Laurentia.

In general, we can say of the complex that it includes rocks of ophiolitic type, that marginal basins, oceanic sea-mounts (and/or islands) and supra-subduction volcanic arcs all contributed to the shunted and crumpled ensemble which was obducted onto the southern margin of the continental Midland Valley terrane. Three phases of amalgamation and uplift have been distinguished. The first occurred at around 576 Myr (i.e. in the late Precambrian), the second towards the end of the Cambrian (*c.*505 Myr) and the final one, between 490 and 470 Myr, i.e. probably in the early Ordovician. All were consequences of the oblique collisional stages in the history of Iapetus. *Fig. 10.7* shows one plausible interpretation suggesting a change from northerly to southerly dipping subduction during the evolution of the complex.

The question of how one can identify the different tectonic settings in which the lavas were erupted will be asked, especially bearing in mind the bruised and battered condition of the rocks. The answer comes from igneous geochemistry. Of the ninety-two naturally occurring elements on this planet, many are present in basaltic magmas although the great majority are present in very small quantities (measurable only in parts per million) and are referred to as trace elements. Although, to a large extent, the chemistry of the lavas was changed during the hydrothermal alteration (or 'low-grade' metamorphism) that affected the rocks, the relative concentrations of some of the trace elements remain fixed. Importantly, these relative concentrations can be diagnostic for magmas erupted in different situations. These have been meticulously determined for modern basalts from known environments and the data can be used to deduce the tectonic settings in which ancient basalts were formed. We have already seen (Chapter 9) how lava chemistry has been employed to reveal the story of the subduction of Iapetus sea floor beneath much of Scotland in Old Red Sandstone times. The extent to which geochemical techniques can be used to 'finger-print' even the most unprepossessing pieces of crushed and recrystallised basalt is a remarkable achievement of modern science.

Obduction commonly gives rise to very considerable changes in relief: in the Ballantrae Complex faulting appears to have created a rugged sea-floor topography in which some of the deeper parts of the lithosphere (namely gabbros and serpentinised peridotites) became exposed on the sea floor as submarine escarpments. The evidence for this comes from blocks of gabbro and serpentinite within some of the volcanically-derived sediments, deduced to have fallen from these faulted escarpments. Some of the serpentinised peridotite fragments have coatings of calcareous algae showing that they must have been exposed on the sea floor prior to their falling to their final resting place. Final closure and the squeezing up of the faulted ophiolite slices, piggy-back fashion, against the continental margin has been dated to an age of around 482 million years.

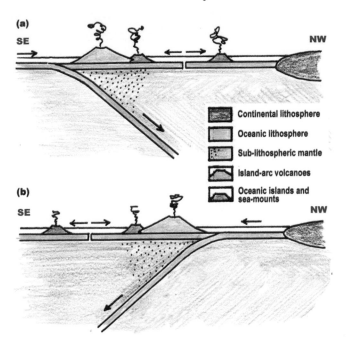

Fig. 10.7 Suggested model for evolution of the Ballantrae Complex. Early subduction towards the North (a) gave way later to southerly subduction (b). Stippled areas in mantle indicate regions in the 'mantle wedge' where magmas were generated to supply overlying island-arc volcanism. *(After Oliver et al., 2002)*

The Southern Uplands

While exposure of the Southern Uplands rocks is superb along much of the coast, inland exposures are poor. Beneath a partial veneer of Devonian and Carboniferous strata, the rocks are of Ordovician and Silurian age and have been affected by Caledonian deformation. Once again the detailed evolution of these older rocks is a matter of vigorous dispute. Most of the visible Ordovician–Silurian rocks originated as sands and muds laid down as trench deposits at a continental margin and it is likely that northward subduction of the Iapetus plate caused the sedimentary pile to be stacked up against the continental crust to the north. One hypothesis proposes that a collision between the latter and a volcanic arc caused compression of an intervening back-arc basin.

Almost all of the volcanic rocks within the Southern Uplands lie within the Ordovician belt that forms the northernmost tract of the Southern Uplands (including the Lammermuir and Moorfoot Hills). Recently, however, ash horizons have been identified in the overlying Silurian strata.

Fig. 10.8 Pillow basalts of early Ordovician age at Downan Point on the south-west coast, south of Ballantrae.

Some of the oldest lavas (pillow basalts) have been described from near Abington, although there are several other occurrences. Beautiful crops of pillow lavas, regarded as part of a former sea-mount, are to be seen in an essentially unsheared condition at Downan Point, on the west coast, south of Ballantrae (*Fig. 10.8*). One of the bigger outcrops of Ordovician volcanic rocks is seen at Bail Hill (near Sanquhar), where the chemistry of the lavas also points to a sea-mount origin. Yet another group of intra-plate lavas (the 'Wrae Volcanics'), which occurs in Tweedale, is remarkable in containing rhyolites suggesting an origin in a large and complex volcano: an oceanic island rather than a mere sea-mount is indicated. In brief, the scattered and generally unsatisfactory outcrops of the Ordovician volcanic rocks in the Southern Uplands present an image of intra-plate volcanoes, fissure volcanoes erupting from the sea floor (back-arc basin or even possibly 'open ocean') and spreading axes as well as others suspected of belonging to supra-subduction arcs.

Latest Precambrian submarine volcanoes of the Highlands

We turn now to some of the volcanism that took place roughly a hundred million years before the early Palaeozoic events considered above. This activity took place very late in Precambrian times and related to the conception and birth of Iapetus rather than to its declining and terminal stages.

The Grampian Highlands, bounded by the Highland Boundary Fault to the south and the Great Glen Fault to the north, are composed of an intensely folded sequence of rocks collectively known as the Dalradian, after the ancient kingdom of Dalriada that covered much of the SW Highlands. Most of the Dalradian rocks started out as marine sediments deposited onto continental margins of varying depths, brought in by rivers from the surrounding lands in the period from 800 to around 590 Myr. The deposition began when Rodinia was a coherent continental mass and ended as it began to rift and come apart.

Long after they were laid down, the Dalradian rocks became caught up in the Caledonian Orogeny and subjected to moderately high pressures and temperatures (up to around 400° C). In the course of this widespread metamorphism the volcanic products (mainly basalts) and associated shallow-crustal intrusions (mainly dolerites) underwent metamorphic changes to both their mineralogy and textures. New minerals formed in this way included greenish, hydrated silicates – members of the amphibole and chlorite mineral groups – grown mainly at the expense of former pyroxenes. Their crystal growth during a regime of directed pressures caused them to adopt parallel orientations, either as needle-like or plate-like crystals. Since the minerals split or cleave more readily in some directions than others, the result of this oriented growth was to produce rocks which split easily as well as having a greenish coloration. Because the geological term for such readily cleavable metamorphic rocks is 'schists', we can describe most of the volcanic rocks caught up in the Caledonian Orogeny as comprising 'green schists'. The cross-sections of individual lava pillows in these sea-floor sequences would originally have been nearly circular, but orogenic deformation has since transformed them into ovoids or lenticles (*Fig. 10.9*) and, in some cases, so distorted the original morphology as to render it unrecognisable.

For a long time Rodinia was subjected to extensional forces stemming from convective movements in the underlying mantle. These caused the faulting and subsidence that gave rise to the marine basin in which the Dalradian rocks accumulated. There were several minor outbreaks of volcanism in the basin producing pillow basalts and associated pyroclastic deposits, subsequently transformed to green schist meta-volcanic rocks that we now see sandwiched among assorted meta-sedimentary rocks. The earliest of these, dated very roughly to 650 Myr, occurs in the eastern Grampians.

A widespread layer, up to several metres thick and rich in the heavy mineral baryte (barium sulphate), occurs in the Dalradian succession and is likely to have been created in one of these magmatic episodes. This baryte deposit, interbedded with other Dalradian meta-sedimentary rocks,

Fig. 10.9 Pillow lavas in the Tayvallich sucession. Deformation in the Caledonian Orogeny has produced the elongate lenticular forms. Coin 25 mm diameter.

has been of great economic importance since the mineral is required for drilling-muds for use in the North Sea oil and gas wells. The Foss Mine has produced some 50,000 tons annually since 1984 from this deposit and minerals of lead, copper and zinc are minor components in this layer. The deposit probably owes its origin to the interaction of seawater with hot fluids of magmatic origin that arose through the shallow crust, in conjunction with the volcanism.

All these relatively insignificant outbreaks of submarine volcanism were a prelude to a much more important volcanic event that took place around 600–595 Myr. This may have been caused by an acceleration in the rate of stretching that promoted rapid basin deepening and which also led to melting in the underlying mantle. The products were the Tayvallich lavas, so-named from their type locality, near Tayvallich in the south-western Highlands. These lavas and intercalated pyroclastic (meta-hyaloclastite) horizons attain a maximum thickness of about 2 km near Loch Awe and along the Tayvallich Peninsula. Whereas most of the Tayvallich succession consists of pillow lavas of submarine origin, there are some indications of subaerial eruption towards the north-east.

The Tayvallich lavas are underlain by an extensive array of doleritic and gabbroic sills representing the magmas that, failing to reach the sea floor, spread laterally along bedding planes in the marine sedimentary rocks. Some, like examples cited from the Siluro-Devonian phenomena (Chapter 9), may have been injected into soft and unconsolidated sediments. The sills are well seen in Knapdale, northern Kintyre and around Loch Awe and the whole assemblage of lavas and sills can be traced laterally for over 100 km. Since the sills have a combined thickness of up to 3 km, the whole succession of extrusive and intrusive rocks could attain a total thickness of around 5 km. To the north and west, the outcrop of the pre-Caledonian Tayvallich lavas disappears beneath the Lorne lavas, erupted about 180 million years later in post-Caledonian times (Chapter 9).

Fissuring of the thinning lithosphere provided channel-ways (dykes) for the Tayvallich magmas to ascend, and dykes on Jura could represent some of the feeders that supplied the sea-floor volcanoes and their underlying sills. Eventually these same extensional forces brought about rupture of the continental lithosphere and the birth of the new ocean. The inference that the extent of mantle melting increased progressively from the start of volcanism to the time of the final parting of the continent comes from the changing chemistry of the Tayvallich lavas upwards through the sequence. Comparisons between the composition of the Tayvallich basalts and more modern basalts, which are known to mark the initial phases of continental break-up, provided support for the conclusion that the Tayvallich erup-tions heralded the opening of Iapetus. The end of Tayvallich volcanism was followed by a quiescent interval during which poorly-sorted, shallow sea sands were deposited. Sea-floor volcanism then resumed and produced another sequence of green schist pillow lavas up to 500 m thick in the Loch Avich area, nearly 20 km SE of Oban.

Up to this time, the only life forms witnessing the Dalradian eruptions would have included worms and other invertebrates. Although these left no tangible fossils, we know of them from the tracks or burrows they left in the sea-floor sediments. Any ancestors of ours in these times would also have been marine, most probably cartilaginous eel-like creatures of which we have little or no detailed knowledge.

We have already seen that more recently (a mere 60 million years ago), extension of the great northern continental mass of North America, Greenland and Europe led to widespread volcanism, of which the Hebridean Volcanic Province is a small part. This Palaeogene volcanism was a prelude to continental separation and the creation of the North Atlantic Ocean, and there are comparisons to be drawn between this and the situation outlined above that culminated in Iapetus. However, in the latter case, to the best of our knowledge, the volcanoes all (or almost all) erupted below sea-level and, at least from the evidence in Scotland, no subaerial lava fields were created.

Chapter 11

Volcanoes seen as through a glass darkly:
the earlier Precambrian record

As L. P. Hartley famously observed: 'The past is a different country. They do things differently there.' We started off this story in a distinctly 'different country' when considering the extraordinary aspect that the Hebridean region would have presented some 60 to 55 million years ago. However, by the time we delve more than ten times further into the past even the most elastic imagination must start to experience the first signs of boggling. We would indeed be in 'a very different country' or rather, a very different world. The geography of the globe would have been utterly different and the flora and fauna were very primitive; the seas were home to algae and soft invertebrate creatures with grotesque forms. The land surfaces were rocky, sandy or salty deserts; in the late Precambrian these might have supported some lichen growth, although algae and bacteria thrived in the rivers, lakes and hot springs. The composition of the atmosphere itself was different and only from around 1,600 million years ago was oxygen an important enough component for photosynthetic plant life to evolve. As noted in Chapter 2, whereas the oldest rocks preserved in Scotland take us back towards 3,000 million years, it is now known that, in other parts of the world, still older formations give hints as to conditions prevailing another 800 million years earlier.

As outlined in Chapter 2 the term 'Precambrian' relates to times preceding that at around 542 Myr, when hard-shelled animals began to appear. The Precambrian itself is broadly subdivided into the Proterozoic and the Archaean. The start of the Proterozoic (showing evidence of primitive lifeforms) is difficult to define, but began at 2,500 or, according to other definitions, 2,600 million years ago. The Archaean is the name given to the earlier division, which embraces the oldest dated rocks surviving on Earth, with ages extending back to round about 3,800 Myr.

The Dalradian rocks of the Grampian Highlands, considered in the previous chapter, overlie continental rocks believed to have been deformed and metamorphosed around 800 to 900 million years ago. However, the

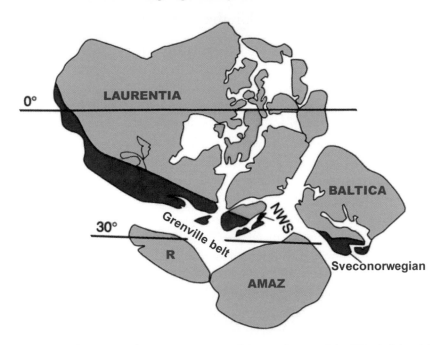

Fig. 11.1 A palaeogeographic reconstruction of the continents of the 'North Atlantic' region at around 725 Myr. The Grenville and Sveconorwegian zone relates to an orogeny that occurred around 1000 to 1100 Myr. R = Rio Plata, AMAZ = Amazonia and NWS = North West Scotland. *(After R. G. Park et al., in The Geology of Scotland 4ᵗʰ edition, N.H. Trewin ed., 2002)*

rocks of this underlying basement are scarcely exposed and little is known of them. When, however, we traverse north across the Great Glen into the Northern Highlands terrane a huge expanse of older rocks is available for study. Between the Great Glen Fault and the low-angled, easterly dipping fault called the Moine thrust (*Fig. 2.1*),]which can be traced from Loch Eriboll SSW to the Sleat Peninsula of Skye, lie the rocks widely referred to as the Moine schists. These are now known to comprise three great groups or packages, probably differing not only in age but also in the geographic environments in which they were deposited, that came to be jammed together during the Caledonian Orogeny. The great majority of the Moine schists appear to have originated as shallow marine sands and muds deposited in rifted continental basins over some millions of years in the period between 1,000 and 850 Myr. These sedimentary formations were predominantly metamorphosed into schists as a consequence of the high temperatures and pressures to which they were subjected during the Caledonian Orogeny.

Evidence of volcanism within the Moines

Within the Moine schists there are some rocks whose mineralogy, chemistry (and very occasionally texture) point to an igneous origin. Some are clearly deformed products of what were originally granites. During the metamorphism of these there was some separation of the darker silicates (hornblendes and biotite micas) from more colourless silicates (feldspar and quartz). The separation resulted in streaky or striped rocks which lack the easily splittable character of schists and are consequently classed as gneisses. Such relatively coarse-grained streaky ('foliated') granitic gneisses underlie, for example, the region of Ardgour in Argyllshire on the northern side of Loch Linnhe (*Fig.11.2*). Since the zircon crystals in these remain essentially unchanged during metamorphism, radiometric zircon dating puts the age of intrusion of the original granite close to 870 Myr. In a book on old volcanoes of Scotland the Ardgour gneiss and other deformed granites of similar age (the West Highland Granite Gneisses) intruded into the Moine schists may be almost wholly irrelevant, since there is absolutely no evidence pointing to their having supplied overlying volcanoes. However, in the following section their possible relevance is spelt out.

There are great numbers of mafic outcrops within the Moine metamorphic terrane. These are basaltic in composition and, in some of the larger and more coarsely crystalline instances, they retain the relict igneous textures of gabbros. Metamorphism of the former basalts, dolerites and gabbros has converted them to rocks in which the principal mineral components are plagioclase feldspar and amphibole. Such rocks are called 'amphibolites'. The temperatures and pressures involved in their formation were somewhat greater than those yielding 'green schists' and the readily splittable (schistose) character of the latter is lost in the amphibolites. The amphibolites occur as lenses or sheets within the Moine meta-sedimentary rocks (specifically those in the Loch Eil region) and almost certainly represent former intrusions of basalt, dolerite or gabbro (dependent on size and length of cooling history). Some were probably sills intruded along the bedding planes of the sedimentary strata. Others, however, whose contacts can be shown to cross-cut the bedding of their country-rocks, were surely dykes. Although dykes may start out as vertically standing bodies, often approximately perpendicular to the strata they have intruded, they will have been rotated into perfect or near perfect parallelism with the layers of their enclosing rocks by the time the whole assemblage has been tightly folded several times. The truth of this can be readily tested by experimenting with plasticine models! Consequently the matter of deciding whether a sheet-like amphibolite in

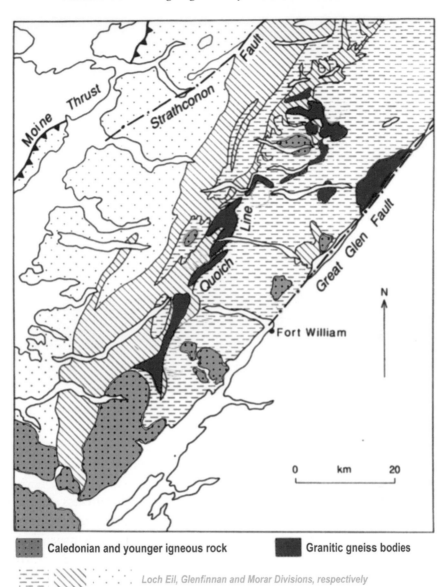

Caledonian and younger igneous rock   Granitic gneiss bodies

Loch Eil, Glenfinnan and Morar Divisions, respectively

Fig. 11.2 Geological map showing distribution of the West Highland Granitic Gneisses in relation o the Great Glen Fault *(After P. E. Brown, The Geology of Scotland, 3rd edition, 1991)*

strongly deformed rocks originally had the form of a dyke or sill or indeed any other form, can be very difficult. This having been said the likelihood is that great numbers of these intrusions constituted a very substantial dyke swarm which has been dated at around 870 Myr, i.e. virtually contemporaneous with the emplacement of the granitic magmas parental to the West

Fig. 11.3 A roadside outcrop in Glen Moriston showing amphibolites (dark layers) in a pale matrix of meta-sedimentary schists.

Highland Granitic Gneisses. That there was not complete contemporaneity is demonstrated by the observation that some of the amphibolite bodies intrude the granite gneisses and are therefore younger. Nonetheless emplacement of the granitic magmas appears to have been followed shortly after (if a period of a few million years can be dismissed as 'short') by massive and regional intrusion of basaltic magmas to form a major swarm of dykes (and accompanying sills etc.). These are very abundant along a narrow NE–SW zone extending roughly from Glenfinnan to Cluanie Lodge on 'the road to the isles' (the A87) and also along a zone close to the northern margins of the Great Glen.

Researches suggest that the basaltic event accompanied extensive lithospheric thinning and crustal rifting. The chemical signature of the bulk of the amphibolites is close to that of basalts generated at mid-ocean spreading centres. Although the intrusions penetrated continental and not oceanic crust the combined evidence of their abundance and their compositions suggests an episode of very significant continental extension over 50 to 200 million years, and which approached the condition of total rupture to produce a new ocean. Allowing the imagination full rein, one may visualise the development of continental rifted basins, extensive melting of the subjacent mantle and copious eruption of basaltic lavas through fissure volcanoes to produce a continental flood basalt province. It has been suggested that heat emanating from basalt magma reservoirs deep in, or just below, the crust in the early stages of this event caused the crustal melting that generated the granite magmas whose ascent narrowly preceded the uprise of the basalt magmas. We have encountered this sort of situation in much more recent geological times as, for example, the production of salic (granitic) magma predating the shallow ascent of basaltic magmas in Rum. In summary, there was a very important magmatic event which took place round about 870 Myr that left its mark in the rocks of the Northern Highlands as well as in the Hebrides (namely in the Ross of Mull). The evidence unfortunately is, as so often, partial and incomplete: it merely hints at the probability of the scenario outlined above and the case must be left non-proven.

The Moine schists, caught up in the multiple phases of Caledonian deformation, were shoved westwards, upwards and outwards, along the Moine thrust to overlie the rocks to the west. It is the trace of the Moine thrust that marks the western boundary of the Caledonian Orogeny. Comparable eastward dipping thrust faults can be found in the eastern USA and also in north-eastern Greenland. Westerly dipping counterparts tracing the eastern margins of the Caledonian Orogeny can be seen in northern Scandinavia. All of these were once parts of a continuous front to the Caledonide mountain belt before their separation by ocean growth brought about by much later, Cainozoic, plate movements (Chapter 5).

Evidence for volcanism in late Precambrian formations west of the Moines

The upward component of the Moine thrust fault was measurable in kilometres while the lateral, outward, motion involved many tens of kilometres. The rocks below and to the west of the shallow-dipping thrust fault are collectively referred to as constituting 'the foreland'. Although over-ridden by the Moine Caledonian rocks, those of the foreland were, to all extents and purposes, left virtually unscathed by the Caledonian mountain-forming events. We see the foreland in a narrow strip exposed along the north-west coast of the Scottish mainland as well as in the Outer Hebrides, Coll, Tiree, Iona and Skye. The foreland consists of ancient metamorphic rocks (the Lewisian basement) with a partially preserved cover of younger Proterozoic and Palaeozoic sedimentary rocks. The Proterozoic rocks are very largely composed of sandstone and, since they are superbly represented in the Torridon Forest area, they are collectively known as the Torridonian sandstones, already referred to in Chapter 6 with regard to the geology of Rum. Canisp, An Teallach, Stac Polly and Suilven are just some of the better known mountains carved from the Torridonian sandstones by glaciers in the past few hundred thousand years.

The Torridonian appears to have been deposited over a long time interval (*c*.1,200–750 Myr). Much must therefore have been contemporaneous with the Moine sedimentation. The Torridonian sandstones are deduced to represent deposits of great eastward-flowing braided rivers reworking earlier sedimentary deposits from the interior of Rodinia. These deposits accumulated in rifted lowland basins that were geographically remote from those of the Moines and covered an undulating landscape that had been eroded out of the much older Lewisian rocks. Whereas the various Moine rocks to the east of the thrust front had experienced such severe metamorphism that the original sedimentary (and occasional igneous) features have largely been erased, the original features of the Torridonian rocks are remarkably well preserved.

The Torridonian sediments, which accumulated to thicknesses of many kilometres, have been subdivided into three units ('groups'). The lowest (and hence oldest) is the Stoer group, so named from its type locality on the Stoer Peninsula, north of Enard Bay. Dated at approximately 1,200 Myr, the Stoer group is of interest here in being the only part of the Torridonian succession from which any record of volcanism has yet been described. A superficial similarity between the Stoer rocks and those of the younger Torridonian groups disguises the fact that they probably had a very different origin. The zone indicated in dark grey on *Fig. 11.1* is that of the 'Grenville–Sveconorwegian Orogenic Belt'. Folding, metamorphism and mountain-building along this belt took place during 1,100–1,000 Myr and it may be thought of as a Precambrian forerunner of the Caledonian Orogeny that took place 500 to 700 million years later. Just as the Caledonian Orogeny attended the demise of an ocean (Iapetus), so the Grenville–Sveconorwegian Orogeny may have been brought about by the closure of an earlier ocean. Little precisely is known of the Grenville–Sveconorwegian events and any ocean that may have preceded them. Whatever the case, the Stoer group rocks formed before the Grenville–Sveconorwegian Orogeny while the younger Torridonian groups post-dated it.

The record of volcanism in the Stoer rocks is sparse, consisting of what are probably volcanic mud flows, perhaps derived from one or two volcanoes in the vicinity, washed down perhaps by desert flash floods. Pumice fragments and small pieces of former volcanic glass in these ancient mud-flows have been severely altered by permeating solutions but were probably basaltic. Although the data are slight and unsatisfactory they may reflect much more extensive volcanism at the time, possibly associated with large-scale faulting to the west, beneath what is now the sea. The evidence is tantalising since evidence of large-scale eruptions in the interval 1,300–1,100 Myr, associated with plate movements and incipient continental break-up, are widespread in North America, Greenland and eastern Scandinavia. Before the opening of the North Atlantic Ocean, the contemporary south Greenland volcanic fields would have lain only a few hundred kilometres at most from Stoer. The name 'Palaeopangaea' has been given to the inferred pre-Grenville continent.

The Rhinnian: possible early Proterozoic volcanic rocks in south-west Scotland

In the last chapter the great sequence of Dalradian meta-sedimentary and meta-volcanic rocks that composes much of the Grampian Highlands was discussed. This sequence was deposited on underlying continental crust which, as we have seen, became increasingly stressed and finally ruptured with continental disintegration. There is little to be seen in Scotland of these sub-Dalradian rocks. Fragments brought up in very fast ascending magmas in the late Carboniferous and Permian events (Chapter 7) offer only tantalising glimpses. Outcrops on Colonsay and SW Islay, however, provide more satisfactory windows through which to view them. These rocks, known as 'the Rhinns', may extend north-eastwards to underlie much of the Dalradian as a substantial crustal unit to the southern side of the Great Glen Fault. The Rhinnian rocks have been correlated with larger tracts of similar nature in Canada, southern Greenland and SW Scandinavia. The Rhinns consist of gneisses whose precursors appear to have been materials, no doubt including both sedimentary and igneous, generated in connection with subduction processes at 1790 to 1800 Myr, along 'an active margin' adjacent to a large, older continental mass. These deductions, reached largely through geochemical investigations, as yet give us only a hint of a situation perhaps like that of the Andean Cordillera of South America where the Pacific plate is subducting eastwards beneath the continent, with an abundance of consequent volcanism.

Evidence of early Proterozoic volcanism in the Loch Maree area

Taking another step back into 'deep time' we may contemplate the scenario hinted at by a group of rocks in NW Scotland in the vicinity of Loch Maree and Gairloch. These are somewhat older than the Rhinnian rocks described above and about seven or eight hundred million years older than the rocks of the Stoer Group. They are components of the Lewisian rocks that form the sub-Torridonian basement. While they are not ancient when compared to some tracts of continental crust elsewhere on Earth, the Lewisian rocks are certainly 'senior citizens' in relation to all other rocks in the British Isles. All, without exception, have been profoundly metamorphosed, generally several times over and are commonly coarsely crystalline gneisses. The complexity – involving folding, fracturing, intrusion and internal recrystallisation of the Lewisian gneisses – was imparted by orogenic events long pre-dating the Caledonian Orogeny.

The Lewisian gneisses outcrop in a broad swathe extending from Cape Wrath, Loch Eriboll and the Butt of Lewis south-south-west to Coll, Tiree and Iona. Most of these rocks acquired their characteristics deep in the crust possibly, as in the case of those of the Scourie district, as deep as 45 or 50 km. Although the nature of these Precambrian orogenies is obscure, the probability is that they too involved some kind of subduction of dense oceanic tectonic plates and the convergence and collision of continental masses. Some of the orogenic events appear to have occurred around 2,900 to 2,700 Myr in the Archaean, while another took place around the Archaean–Proterozoic boundary. The youngest, involving for example, the Rhinnian rocks, occurred approximately 1,900 to 1,500 Myr. The unsurprising result of these repeated Precambrian orogenic experiences is that it is generally extremely difficult to make out what the original rocks were before their metamorphism. Commonly it becomes a matter of determining the bulk chemical composition of the rocks and then making generalised matches with those of unmetamorphosed rocks. Clearly this involves the assumption that the gneisses represent recrystallised products of familiar sedimentary rock-types like sandstones and shales or igneous rocks like granites and basalts. We now know, however, that this is not always so and that there were some sedimentary deposits (e.g. 'banded ironstones') and magmatic products (e.g. the extremely magnesian lavas called komatiites) that tended to be restricted to the very different Precambrian environments.

Despite the general lack of any certainly as to the origins of most Lewisian rocks, there are some tracts, for example in South Harris, Iona and near Loch Maree, which retain characteristics indicating that they originally formed on the surface as sedimentary and/or volcanic strata. Doubtless these subsequently became deeply buried at the time or times of their metamorphism(s), but ultimately they were returned to shallow levels and now outcrop at the surface. It is the Loch Maree group, seen in two areas (around Gairloch and also north-east of Loch Maree) that will be considered here. Although these rocks have been investigated by numerous workers, the account below is based on a recent interpretation by Graham Park and his collaborators.

The Loch Maree group includes rocks of extreme compositions that identify them beyond dispute as metamorphosed sediments. These include marbles (recrystallised limestones), black, graphitic schists (from carbon-rich muds) and black and white banded ironstones, where the black layers are composed of the iron oxide magnetite and the white layers of quartz (originally microcrystalline chert). Much more abundant are schists, whose compositions suggest an origin as greywackes, the rather heterogeneous sandstones we have already encountered as a type of rock abundant in the lower Palaeozoic strata of the Southern Uplands and as the precursor for much of the schist in the Grampian Highlands. In the present instance the greywackes are thought to represent sediment washed into the sea from a volcanic arc overlying a subduction zone. The sedimentary assemblage – involving marbles which are likely to have formed as accumulations of calcareous algae, and the 'banded ironstones' – is of a kind representing relatively shallow marine conditions. The metamorphosed sediments are interlayered with tough dark coloured amphibolites. Analyses of the latter show them to have bulk compositions comparable to those of basalt.

Detailed geochemical forensic work suggests that some of the Loch Maree group amphibolites match basalts erupted in the early stages of island-arc development (and usually not seen above sea-level), tending to confirm the belief that subduction was involved in their genesis. This adds to the evidence that subduction, a process occurring in our own time and promoting eruptions such as those on Montserrat in the Caribbean and Pinatobu in the Philippines, was already well established some two thousand million years ago.

A second category of amphibolites, however, shows unusual trace-element signatures that most closely relate them to oceanic basaltic plateaux. Up until now we have not encountered such oceanic basalts in any of the younger volcanic situations. As we have seen, basaltic magmas in the oceans generally tend to be one of two types: those produced along the mid-ocean spreading centres and those erupted through intra-plate volcanoes that form sea-mounts or, if big enough, oceanic islands like those of Hawaii. As explored in Chapter 10, faulted slices of both kinds are present in the early Palaeozoic rocks of the Ballantrae region. However, 'chemical finger-printing' of the second category of amphibolites in the Loch Maree group ally them to the great submarine volcanic plateaux which have only relatively recently been identified beneath present-day oceans. The greatest of these is the Ontong–Java plateau, a vast region of magmatic outpouring in the SW Pacific Ocean, north-east of New Guinea and Australia. The submarine Ontong–Java eruptions took place on a gigantic scale in the Cretaceous, i.e. in recent times compared to those of the early Proterozoic under considera-tion, and are inferred to relate to the uprise of a large-scale mantle plume. Thus, although the data are very tentative, one may speculate that some of the amphibolites in the Gairloch–Loch Maree area may represent bits of an ancient oceanic plateau comparable to 'modern' examples like the Ontong–Java plateau. These second category amphibolites occur in conjunction with some schists considered to have originated as deep-ocean floor sediments, produced in a very different environment from the other Loch Maree meta-sedimentary rocks indicative of shallower marine deposition.

Some coarser-grained, more siliceous igneous rocks constitute yet another component of the Loch Maree group. These, the youngest rocks in the group, have intrusive relationships to the Loch Maree group meta-sedimen-tary rocks and are rather precisely dated at 1,900 Myr. These intrusions, with compositions intermediate between diorites and granites (but rather closer to granites), have also been metamorphosed to form what are called the Ard gneisses. Their chemical characteristics resemble those of supra-subduction zone magmas. Whether the magmas that formed the precursors to the Ard gneisses ever erupted at the surface is unknowable but had they done so their products would have been comparable to andesites and dacites. The Ard gneisses are tentatively construed as providing further evidence for the exist-ence of a volcanic arc related to subduction of an oceanic plate.

According to some recent investigation, the subduction history inferred from the Loch Maree group led ultimately to closure of an ocean as two continental blocks were dragged into collision. If correct, we are thinking in terms of an ocean far more ancient than the one that may have led to the Grenville–Sveconorwegian Orogeny. In this interpretation the Loch Maree group represents rock formations caught between the jaws of a closing vice, these being represented by the more ancient striped gneisses that bound the Loch Maree group. All these rocks, bounding gneisses and intervening meta-volcanic and meta-sedimentary rocks, were fated to experience further deformation and recrystallisation in Earth movements (the 'Laxfordian events') between 1,900 and 1,500 Myr. Although the supposition is that most of the amphibolites would formerly have been submarine pillow lavas it comes as no surprise that, in such horrendously maltreated rocks, the primary volcanic features have been erased. However, it should not be thought that the unrecognisability of these 'once-upon-a-time' lavas is simply a function of their comparatively great age. In many parts of the world volcanic rocks of comparable, and indeed much greater, age have survived the exigencies of time and splendidly retain their original surface features. For example, perfect pillowed morphologies can be seen in submarine lavas in southern Greenland, whose age is similar to that of the Loch Maree amphibolites.

If an ancient ocean *was* consumed during the closure of two continent masses as postulated in the above reconstruction, the question may be asked 'What happened to all the rest of the oceanic rocks, namely the oceanic lithospheric mantle and overlying crust of intrusive and extrusive rocks?'. The answer may be that all of these became cold and dense and sank for recycling within the deep mantle. We can regard this as the normal procedure and remember that it is only in abnormal mechanistic accidents (obduction) that some fragments of the oceanic lithosphere get thrust up and over the continental edges to produce an ophiolitic assemblage, as happened in the Ordovician. However, if we accept the above explanation of the evolution of the Loch Maree group, this assemblage of meta-sedimentary rocks, 'primitive arc' lavas, supposed oceanic plateau lavas and intrusive rocks escaped subduction because of its greater buoyancy than the older mafic and ultramafic rocks. To give a simplistic analogy, they behaved more like corks bobbing in a rain gutter, too buoyant to go down the drain!

As we have seen earlier, palaeomagnetic data indicate that the various continental fragments that came together over the millennia to form what is now celebrated as Scotland have mostly had a generalised northward motion. Thus we know that the late Carboniferous rocks of Scotland formed in an equatorial environment whereas those produced in the Ordovician originated in southern latitudes and the Stoer group rocks, which date from roughly 1,200 Myr, may possibly have accumulated in Antarctic latitudes. But when we contemplate the early Proterozoic rocks like those of the Loch Maree group all sense of their global positioning is effectively lost. In the recent geological past the tectonic plates have been (and are) moving at rates commonly between 1 and 10 cm per year. Even if we took an average annual motion of 5 cm, the 2,000 million years since the Loch Maree group rocks formed could give them an itinerary of 100 000 km. In the Precambrian at such a time, plate motions may well have been faster and altogether more chaotic, and we have to conclude reluctantly that it is unlikely that we shall ever know in what part of the world they originated.

Dykes in the Lewisian suggesting early Proterozoic flood basalt volcanism

The Lewisian basement is intruded by the Scourie dyke swarm, so-called on account of the abundance of these dykes in the vicinity of Scourie (*Fig. 11.4*). The typical trends are from NW–SE to E–W and the dykes range up to over 100 m wide. Radiometric dating suggests ages between 2,400 and 2,000 million years. Across much of the Lewisian terrane the Scourie dykes have been severely affected by shearing and folding that took place several hundred million years later, between 1,700 and 1,500 Myr, in what are referred to as the Laxfordian event(s). Although many of the dykes can be described as dolerites grading, in the larger, slower cooled examples to gabbros, some are highly magnesian and comprise ultramafic rather than mafic rocks. All the dykes, even those unaffected by the Laxfordian defor-mations, have undergone some degree of metamorphism (*Fig. 11.5*).

Despite the possibility of an overlap in ages between the younger Scourie dykes and the Loch Maree meta-volcanic rocks, no genetic relationship between them has been recognised. The dykes present evidence that they were emplaced dilatationally, i.e. that the wall rocks were being pulled apart to admit the magmas, and the swarm was clearly emplaced during times of very substantial crustal extension. In some cases, and perhaps generally, the dyke walls were also moving sideways with respect to each other. It has been demonstrated that some at least of the Scourie dykes were emplaced during right-lateral shearing. Much remains to be learnt about this remarkable and enigmatic swarm, with major questions as to whether there was more than one intrusive event and what was the total time involved. It is, however, clear from determination of the temperatures and pressures involved in the forma-tion of the crystal assemblages and crystal fabrics, that the dyke rocks that we now observe at the surface acquired their characteristics at depths of 10 to 20 km in the crust. In other words profound uplift took place subsequent to their formation, with up to 20 km of overlying crust being stripped off by erosion to create the hilly landscape on which the Torridonian sediments accumulated.

There can be no doubt that we see only a tiny remanent of the original Scourie dyke swarm. The westerly extensions of the dykes mostly lie

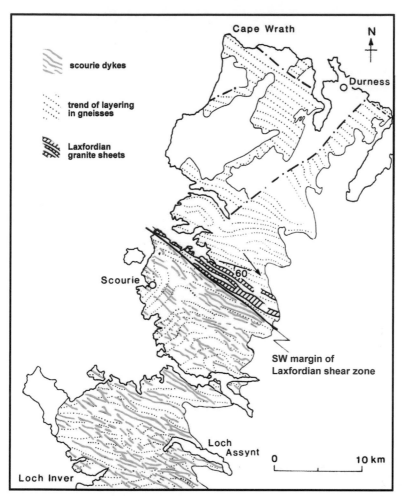

Fig. 11.4 Map of the Scourie dyke swarm in the region between Scourie and Loch Inver. *(After R. G. Park et al., Geology of Scotland, N.W. Trewin ed., 2002)*

unseen on the continental shelf while their easterly extensions are either buried beneath younger formations (Torridonian sandstones and overlying Cambrian–Ordovician sedimentary rocks) or they have been over-ridden by the Caledonian Moine thrust. Nonetheless, their abundance and thicknesses point to the emplacement of huge volumes of basaltic (and more magnesian) magmas probably amounting to hundreds, if not thousands, of cubic kilometres. We must accordingly contemplate a time of profound continental extension accompanied by lateral shearing during which great volumes of magma were sweated out of the mantle. Only a small proportion of dykes will ever have sufficient thermal energy to make the whole traverse of the crust to erupt at the surface. Nevertheless the Scourie swarm

Fig. 11.5 Three very small members of the Scourie Dyke Swarm, unaffected by the Laxfordian events, cutting gneisses on South Uist, Outer Hebrides.

is of such magnitude that it is almost inconceivable that it did not have volcanic expression, giving vent to thick and widespread sequences of 'flood basalts' on a prodigious scale.

If we try to picture what the geographic environment was like at the time, a considerable act of imagination is required since we are dealing with an early Proterozoic scene about half as old as the planet itself. To do so, however, does not require conjuring up a fictional land like Lewis Caroll's in *Through the Looking-Glass*, C. S. Lewis's Narnia or Tolkien's Middle Earth. Even though we cannot be precise about the details, we can discern the rough outlines. We can, for instance, be certain that continents and oceans existed, that there were mountains and plains and that the waves crashed on rocky coasts and creamed smoothly up sandy shores. The rivers swelled in spate in the rainy seasons and the landscape parched during the droughts.

The fact that the Scourie dykes were intruded into a continent that was experiencing tension and shearing stress allows us to envisage a rifted volcanic region. On the assumption that eruptions were sub-aerial – rather than on a continental shelf beneath shallow seas – they would have seen basalt lavas spilled across a fissured landscape on a much greater scale than for example in the Palaeogene. It would certainly have been a desert landscape devoid of vegetation. The Precambrian was not merely 'a different country' but one in which things tended to be done 'bigger and better'. The young Earth simply possessed more energy. The compositions of the Scourie dykes imply rather elevated degrees of mantle melting while the widths of many may be taken to imply very energetic dilatation (high rates of release of strain). One or more mantle plumes may have been implicated

Fig. 11.6 Lenticles of amphibolitised mafic or ultramafic rock within pale gneisses, South Uist, Outer Hebides. Coin 25 mm diameter.

as a heat source. The concentration of dykes in some areas, e.g. approaching thirty percent around Gairloch, suggests a dramatically stretched crust that was approaching complete sundering and the birth of a new ocean. If indeed this happened, the evidence is long gone.

Any speculation as to the nature of plate tectonics and volcanism attending the Scourie dyke swarm is far better grounded than any concerning the Lewisian country-rocks into which they were intruded. In these latter there are many lenses of mafic rocks which are presumed to have been basaltic lavas or intrusive dolerites or gabbros (*Fig. 11.6*). It is impossible, at least on present information, to decide whether these were parts of continental crust or whether they originated in oceanic crust which subsequently became incorporated into continental crust. Ultramafic materials, which probably commenced their existence as peridotites, are also commonly encountered. But whether these presumed peridotites represent remnants of extremely magnesian lavas (komatiites), concentrates of olivines and pyroxenes in former great mafic magma chambers, or whether they represent slices of lithospheric mantle that have become detached from their sources and stirred into the present continental cocktail presents an intractable problem. The more feldspar-rich and generally more silica-rich gneisses within which these mafic and ultramafic slivers occur often have bulk chemical compositions much like those of andesitic volcanoes associated with subducting oceanic crust but also much like those of the sediments derived from the erosional decay of such volcanoes. So we are presented with mere shadows of a long-gone Precambrian world in which the oceans waxed and waned, the continents grew by processes involving subduction and in which some forms of intra-plate volcanism took place within both the oceans and the continents.

Largo Law, Fife: the eroded stump of one of the multitude of late Palaeozoic intra-plate basaltic volcanoes in the Midland Valley of Scotland.

Chapter 12

Epilogue

It is perhaps fitting to end this account with a brief overview of the volcanic history of Scotland. Extensional stresses acting on continental crust produced the early Cainozoic dyke swarms of the Hebridean Province and, in the Skye swarm, approached the pull-apart, new ocean-forming, stage. This, however, did not happen there but failure of the crust and separation to form the start of the North Atlantic Ocean occurred further west, beyond Rockall. A sudden, short-lived and quite dramatic extensional event had taken place some 240 million years earlier, close to the Carboniferous–Permian boundary, with the production of the broad roughly E–W dolerite dykes, although as has been explained, these probably did not give surface volcanoes but instead fed voluminous shallow crustal sills. It was in the late Precambrian where we see in the Tayvallich volcanics of Argyllshire lavas erupted on the sea floor on a massive scale, apparently marking the opening phases of the Iapetus Ocean. We then have to go back far into the Precambrian, to the dyke event affecting the Moines at about 870 Myr and still further back to over 2,000 Myr, to the Scourie dykes, to find evidence of other major continental dyke swarms. It is by no means beyond possibility that ocean opening attended these Precambrian events but, if it did, the chances are that all evidence was subsequently devoured by subduction culminating in continental collision.

The youngest Scottish supra-subduction volcanoes were those of the late Silurian–early Devonian. These were the 'Old Red Sandstone' volcanoes, whose eroded remnants include major landscape features such as Cheviot, Glen Coe and Ben Nevis. This subduction was still occurring after Iapetus had closed and the volcanism took place in a continental setting for which the volcanic ranges of the Caucasus and Elbruz may provide approximate modern analogues. However, the scraps of left-over volcanic rocks relating to earlier (Ordovician) stages of Iapetus, particularly those of the Ballantrae region, represent a junk-heap of assorted island-arc volcanoes, back-arc sea-floor lavas and intra-plate oceanic volcanoes. As we have seen, there are hints of subduction-related volcanic arcs, together with a possibly oceanic volcanic plateau, in the Loch Maree meta-volcanic and meta-sedimentary assemblages.

Classifications are, of course, a human artefact and natural phenomena are often only awkwardly pigeon-holed. Thus the Hebridean Palaeogene volcanoes can be regarded as 'continental intra-plate', heralding the break-up of Scotland–Norway from eastern Greenland, with the energy source probably provided by the upwelling of a deep mantle plume. The magmatism of the Carboniferous–Permian is more typical of continental intra-plate activity in which, for the most part, there is little reason to invoke mantle plume activity. The magmas that fed the hundreds of small volcanoes across south and central Scotland more probably owed their existence to pressure relief melting of the mantle associated with fault movements that were hundreds of kilometres north of contemporaneous ocean closures and mountain building.

I am sometimes asked if Scotland's volcanoes are completely extinct or merely dormant. I have no doubt that they have exceeded their natural shelf-life and will never erupt again. This said, we live on a vibrant planet with no shortage of internal energy. The plates still move and the ocean floors are not static. The chances are that the North Atlantic, now steadily expanding, will go into reverse; the continental margins will converge and collision will ultimately ensue. Volcanoes will play their part in all these processes. But before holding one's breath one may relax in the anticipation that any such closure may be at least fifty to a hundred million years into the future.

In conclusion, igneous rocks erupted from volcanoes or formed as intrusions beneath them played a major role in building up the whole fascinating collage of terranes that now constitute Scotland. Travels through the beautiful landscapes thus provide no shortage of food for thought and contemplation, reminders that the awful powers of destruction vested in volcanoes were also forces for growth. Erosion of volcanoes and their roots produces sediments and it has been observed that sedimentary rocks are to igneous rocks what sawdust is to forest trees. Taking this viewpoint one could conclude that Scotland has ultimately grown from repeated magmatic episodes over as far back in time as we can read the geological record. Many a heather-covered hillside is a gravestone to a once awe-inspiring volcano.

Select Bibliography

1 The Midland Valley: British Regional Geology, 3rd edition (eds. I. B. Cameron and D. Stephenson). The British Geological Survey, 1985.

2 The Grampian Highlands: British Regional Geology, 4th edition (eds. D. Stephenson and D. Gould). The British Geological Survey, 1995.

3 Igneous Rocks of the British Isles (ed. D. S. Sutherland). John Wiley & Sons, 1982.

4 Geology of Scotland, 3rd editon (ed. C. Y. Craig). The Geological Society, 1991.

5 The Geology of Scotland (ed. N. H. Trewin). The Geological Society, 2002.

6 Carboniferous and Permian Igneous Rocks of Great Britain (eds. D. Stephenson, S. C. Loughlin, D. Millward, C. N. Waters and I. T. Williams). Geological Conservation Review Series. Joint Natural Conservation Committee & British Geological Survey, 2003.

7 Caledonian igneous rocks of Great Britain (eds. D. Stephenson, R. E. Bevins, A. J. Highton, D. Millward, I. Parsons, P. Stone and W. J. Wadsworth). Geological Conservation Review Series. Joint Natural Conservation Committee and British Geological Survey, 2002.

Index of Place Names

Note: page numbers in **bold** denote illustrations.

Index of Selected Technical Terms

Note: page numbers in **bold** denote illustrations.